Agriculture Issues and Policies

Agriculture Issues and Policies

Pomegranate: For Horticulture Students and Farmers
Arkalgud Nanjundaiah Ganeshamurthy, PhD (Editor)
2023. ISBN: 979-8-88697-812-4 (eBook)

Research Advancements in Organic Farming
Jatindra Nath Bhakta, PhD (Editor)
Sukanta Rana, PhD (Editor)
2023. ISBN: 979-8-88697-519-2 (Hardcover)
2023. ISBN: 979-8-88697-580-2 (eBook)

Legumes: Nutritional Value, Health Benefits and Management
Phetole Mangena, PhD (Editor)
2023. ISBN: 979-8-88697-558-1 (Softcover)
2023. ISBN: 979-8-88697-583-3 (eBook)

Jute: Cultivation, Properties and Uses
Matthieu Issa (Editor)
2022. ISBN: 979-8-88697-490-4 (Softcover)
2022. ISBN: 979-8-88697-505-5 (eBook)

Pistachios: Cultivation, Production and Consumption
Shaziya Haseeb Siddiqui, PhD (Editor)
Shoaib Alam Siddiqui, PhD (Editor)
2022. ISBN: 978-1-68507-949-9 (Hardcover)
2022. ISBN: 979-8-88697-063-0 (eBook)

More information about this series can be found at
https://novapublishers.com/product-category/series/agriculture-issues-and-
policies/

Philip L. Bevis
Editor

Biofertilizers

Agricultural Uses, Management and Environmental Effects

Copyright © 2023 by Nova Science Publishers, Inc.

All rights reserved. No part of this book may be reproduced, stored in a retrieval system or transmitted in any form or by any means: electronic, electrostatic, magnetic, tape, mechanical photocopying, recording or otherwise without the written permission of the Publisher.

We have partnered with Copyright Clearance Center to make it easy for you to obtain permissions to reuse content from this publication. Please visit copyright.com and search by Title, ISBN, or ISSN.

For further questions about using the service on copyright.com, please contact:

Copyright Clearance Center
Phone: +1-(978) 750-8400   Fax: +1-(978) 750-4470   E-mail: info@copyright.com

NOTICE TO THE READER

The Publisher has taken reasonable care in the preparation of this book but makes no expressed or implied warranty of any kind and assumes no responsibility for any errors or omissions. No liability is assumed for incidental or consequential damages in connection with or arising out of information contained in this book. The Publisher shall not be liable for any special, consequential, or exemplary damages resulting, in whole or in part, from the readers' use of, or reliance upon, this material. Any parts of this book based on government reports are so indicated and copyright is claimed for those parts to the extent applicable to compilations of such works.

Independent verification should be sought for any data, advice or recommendations contained in this book. In addition, no responsibility is assumed by the Publisher for any injury and/or damage to persons or property arising from any methods, products, instructions, ideas or otherwise contained in this publication.

This publication is designed to provide accurate and authoritative information with regards to the subject matter covered herein. It is sold with the clear understanding that the Publisher is not engaged in rendering legal or any other professional services. If legal or any other expert assistance is required, the services of a competent person should be sought. FROM A DECLARATION OF PARTICIPANTS JOINTLY ADOPTED BY A COMMITTEE OF THE AMERICAN BAR ASSOCIATION AND A COMMITTEE OF PUBLISHERS.

Library of Congress Cataloging-in-Publication Data

ISBN: 979-8-89113-082-1

Published by Nova Science Publishers, Inc. † New York

Contents

Preface		 vii
Chapter 1	**Arbuscular Mycorrhizal Fungi as Biofertilizers for Sustainable Agriculture** 1	
	Praveen Ranadev, Ashwin Revanna, Joanna Dames and Davis Joseph Bagyaraj	
Chapter 2	**Zinc Solubilizing Bacteria: An Emerging Biofertilizer for Sustainable Agriculture** 33	
	Mohammad Reza Sarikhani, Bahman Khoshru and Md Shafiul Islam Rion	
Chapter 3	**The Potential of *Burkholderia* sp. in Meeting the Goals of Sustainable Agriculture** 63	
	Richa Raghuwanshi, Seema Devi and Surya Prakash Dube	
Chapter 4	**Phosphorus Fertility Management in Field Crop Production**..93	
	Mohammad Mirzaei Heydari and Davey L. Jones	
Chapter 5	**The Environmental Significance of Biotechnologically Treated Insoluble Phosphates and P-Solubilizing Microorganisms**........117	
	Maria Vassileva, Eligio Malusa, Vanessa Martos, Luis F. García del Moral, Stefano Mocali, Loredana Canfora, Giacomo di Benedetto, Aspasia Lykoudi, Pedro Cartujo and Nikolay Vassilev	

vi Contents

Chapter 6 **A Sustainable Approach to Increase the Bioavailability of Iron in Plants: The Potential of Iron-Solubilizing Microbes**..............139
Bahman Khoshru and Mohammad Reza Sarikhani

Chapter 7 ***Azotobacter* as a Biofertilizer in Sustainable Agriculture** 159
Mohammad Reza Sarikhani and Mitra Ebrahimi

Chapter 8 **The Role of Biofertilizers in Sustainable Agriculture** 179
Bhupinder Dhir

Index 203

Preface

This book contains eight chapters that detail biofertilizers. Chapter One discusses arbuscular mycorrhizal (AM) fungi, known to form symbiotic associations with most of the crop plants important in agriculture, horticulture and forestry. Chapter Two critically examines the current state-of-art use of zinc solubilizing bacteria strains as biofertilizers and the important roles performed by these beneficial microbes in maintaining soil fertility and enhancing crop productivity. Chapter Three presents Burkholderia, an important bacterial species that directly promotes plant growth. Chapter Four investigates the role of different phosphorus sources on phosphorus bioavailability and explores its cycling in both plants and soil, while also exploring its effects on plant growth and agricultural production. Chapter Five describes the multifaceted environmental impact of microbially treated insoluble inorganic natural sources. Chapter Six discusses the importance of iron (Fe) in soil and plants, the factors affecting its solubility and deficiency in the soil, and the potential of microbial iron solubilizers as a sustainable approach to enhancing Fe availability in soil, and promoting plant growth. Chapter Seven focuses on Azotobacters and assesses their diversity, action mechanisms, ecological significance and its biotechnological applications. Lastly, Chapter Eight discusses the role of biofertilizers in sustainable agriculture.

Chapter 1

Arbuscular Mycorrhizal Fungi as Biofertilizers for Sustainable Agriculture

Praveen Ranadev[1]
Ashwin Revanna[1]
Joanna Dames[2]
and Davis Joseph Bagyaraj[1]

[1]Centre for Natural Biological Resources and Community Development (CNBRCD)
41 RBI Colony, Anand Nagar, Bengaluru, India
[2]Rhodes University, Makhanda (Gramhamstown), South Africa

Abstract

In intensive agriculture, chemical fertilizers are applied on a large scale to increase crop production to meet the needs of the increasing population. The excessive use of chemical fertilizers resulting in adverse effects on soil health and the environment is now well understood. The current day emphasis is on sustainable agriculture which uses less of these chemical inputs. Biofertilizers or microbial inoculants play an important role in sustainable agriculture. Biofertilizers contain living microorganisms when applied to seed, plant surface or soil promote plant growth by increasing the supply or availability of nutrients to the host and also increase soil fertility. The common biofertilizers used are microorganisms fixing atmospheric nitrogen, mobilizing phosphorus, mobilizing potassium and plant growth-promoting rhizomicroorganisms. Plants do form symbiotic relationships with some of these microorganisms. Arbuscular mycorrhizal (AM) fungi are known to form symbiotic associations with most of the crop plants important in agriculture, horticulture and forestry. AM fungi are non-septate Glomeromycetous fungi. These fungi are obligate symbionts and have

In: Biofertilizers
Editor: Philip L. Bevis
ISBN: 979-8-89113-082-1
© 2023 Nova Science Publishers, Inc.

not been cultured on nutrient media. These endophytes are not host-specific although they exhibit host preference. Several investigations indicated that even in unsterile soils under field conditions plants respond to inoculation with efficient strains of AM fungi. The mechanism of improved plant growth caused by mycorrhizal inoculation has been investigated by many workers. Greater soil exploration by mycorrhizal roots and the associated hyphal network as a means of increasing phosphate uptake is well established. They also improve the uptake of other diffusion-limited nutrients like Zn and Cu. The other beneficial effects are their role in biological control of root pathogens, hormone production and greater ability to withstand drought and other abiotic stresses. Many high value crops are grown in nursery beds/root trainers or micro-propagated and then planted in the main field. AM fungal inoculation under these conditions brought out increased plant growth not only in the nursery but also after transplanted in the main field. Soil application of AM fungi has been shown to improve the growth of directly field sown crops with up to a 50% reduction in P fertilizer application. Co-inoculation of AM fungi with other beneficial soil microorganisms is more useful in improving plant growth thus suggesting the development of suitable microbial consortia for inoculating different crop plants and thus saving 50% of NPK fertilizers. Different aspects of AM fungi *viz.*, their taxonomic position, isolation and maintenance, mycorrhizal dependency and selection of efficient AM fungal strains for inoculation, plant growth response to inoculation and mechanisms involved, inoculum production, application methods, inoculation of AM fungi with other beneficial soil organisms, quality assessment of AM fungal inoculum and future directions of work are covered in this chapter.

Keywords: AM fungi, *Glomus*, biofertilizers, application of AMF, endosymbionts

Introduction

Agriculture is an essential activity for the survival and well-being of humans, as it provides crucial resources such as food, feed, fibre and fuel. However, modern farming methods have led to negative environmental consequences such as soil degradation, nutrient depletion, and greenhouse gas emissions. With the global population increasing and food demand on the rise, it has become crucial to develop sustainable agricultural practices that can meet these demands without harming the environment. One of the key challenges facing modern agriculture is to enhance crop productivity while reducing the negative impacts of conventional farming practices on the environment (Colizzi et al., 2020). To address this challenge, the use of biofertilizers has emerged as a viable alternative to synthetic fertilizers. Sustainable agriculture is an ever-growing field that aims to meet the world's food needs while preserving natural resources and the environment (Mahanty et al., 2017).

Biofertilizers are biological products that contain living microorganisms. When applied to seeds, plant surfaces or soil, they promote plant growth and increase soil fertility. They are an essential component of sustainable agriculture inputs that can boost crop yields and decrease environmental pollution. These microorganisms, such as bacteria, fungi and algae establish beneficial associations with plants. They can fix atmospheric nitrogen, solubilize phosphorus, enhance soil microbial activity and improve plant health and growth (Mahanty et al., 2017). Among the different types of biofertilizers, AM fungi have garnered attention as a promising option for sustainable agriculture. AM fungi are soil-inhabiting fungi that form mutualistic symbiotic relationships with the roots of most plants. These fungi create structures called arbuscules and vesicles inside the root cells, facilitating the exchange of nutrients and other compounds between the plant and the fungus (Bagyaraj et al., 2022a).

Using AM fungi as biofertilizers is a promising approach for sustainable agriculture, offering multiple benefits. AM fungi can increase plant growth and yield, enhance nutrient uptake, and improve soil fertility, among other advantages (Fall et al., 2022). Additionally, they can help to alleviate the negative impacts of environmental stressors, such as drought, salinity and heavy metal toxicity (Garg et al., 2017). AM fungi can significantly improve the uptake of nutrients with limited diffusion capacity, such as phosphorus (P), zinc (Zn) and copper (Cu). Moreover, they can contribute to biological control of root pathogens, and hormone production (Bagyaraj, 2011). When used in conjunction with other beneficial microorganisms such as nitrogen fixers, P solubilizers, and plant growth-promoting rhizomicroorganisms (PGPR), AM fungi can have synergistic effects (Dames and Ridsdale, 2012). In tropical soils, where soil fertility is often low and phosphorus deficiency is prevalent, the benefits of AM fungi are particularly significant. The use of chemical fertilizers can lead to the accumulation of fixed phosphorus in the soil, resulting in reduced effectiveness over time. Therefore, the use of AM fungi as biofertilizers can help to reduce the need for chemical fertilizers and improve the sustainability of agriculture (Mythra et al., 2022; Bagyaraj et al., 2022a).

Over the past few years, there has been a growing interest in using AM fungi as biofertilizers, and numerous studies have explored the mechanisms underlying their beneficial effects on plant growth and soil health. In this article, we provide a comprehensive review of the current knowledge on the use of AM fungi as biofertilizers for sustainable agriculture. Our coverage will include the basic biology and ecology of AM fungi, the mechanisms behind their beneficial effects on plant growth and the practical applications of AM fungi in crop production. Additionally, we will discuss the challenges and limitations of using AM fungi and the potential for future research to optimize their use in sustainable agriculture.

The Fungi

The term "mycorrhiza," which means "fungus root," was first coined by German botanist Frank in 1885 to describe the mutualistic association between the roots of higher plants and certain fungi. These associations are classified based on morphological and anatomical characteristics as ectomycorrhizae and endomycorrhizae. Ectomycorrhizae are common among temperate forest tree species and some can be cultured on laboratory media. Endomycorrhizae include arbutoid, monotropoid, ericoid, orchid and arbuscular mycorrhizal forms (Singh et al., 2020). Arbuscular mycorrhizal (AM) fungi are a type of fungi that form symbiotic associations with the roots of most plants, including economically important crop plants. The name "arbuscular" comes from the characteristic branched structure of the fungal hyphae, which form inside plant roots while the hyphae ramify in the surrounding soil. AM fungi are ubiquitous in soil environments, and they can be found in a range of soil types, from desert soils to agricultural soils. They are also able to form associations with a wide variety of plant species, including many crop plants, making them a key player in soil nutrient cycling and plant productivity. This association began over 400 million years ago with the first land plants, and vascular plants and mycorrhizal fungi have continued coevolving to the present day (Ahmad et al., 2020). This mutualistic relationship between plants and AM fungi plays an important role in nutrient cycling and ecosystem function, particularly in nutrient-poor soils (Bagyaraj et al., 2022a).

Arbuscular mycorrhizal fungi cannot be grown on synthetic media as they are obligate symbionts. They enter plant roots through the root hairs or epidermal cells and grow intracellularly and/or intercellularly in the root cortex. Arbuscules, finely branched haustoria-like structures that facilitate nutrient exchange between the fungus and host roots, develop within the cortical cells. Vesicles of various shapes and sizes, which function as storage organs, are formed in the cortical cells. The presence of vesicles and arbuscules is the defining characteristic of AM fungi in roots (Bucher et al., 2014). The phylum Glomeromycota contains AM fungi, in class Glomeromycetes with four orders including Glomerales, Diversisporales, Paraglomerales and Archaeosporales. There are 11 families, 25 genera and 346 species of AM fungi (http://www.amf-phylogeny.com/ accessed as on May 08, 2023). The commonly occurring genera of AM fungi are *Glomus*, *Rhizophagus*, *Gigaspora*, *Scutellospora*, *Acaulospora* and *Entrophospora* (Bagyaraj et al., 2022a).

AM fungi form a mutually beneficial relationship with a large majority of vascular plants (approximately 90%), but there are certain plant families that rarely associate with them, such as Caryophyllaceae, Brassicaceae, Chenopodiaceae and Cyperaceae. These fungi are widely distributed and occur in plants grown in different regions and environments, including arctic, temperate, and tropical regions, as well as sand dunes, mine spoils, and aquatic habitats (Bagyaraj, 2011). They have the largest host range among all mycorrhizal associations and have been observed in over 1000 plant genera, representing 200 families, with at least 300,000 potential hosts worldwide. Due to the significant overlap in host range among AM fungi, some fungi may have access to thousands of hosts, which is much higher than the estimated number of potential partners if the hosts were evenly distributed among the fungi with no overlap in host range (Bagyaraj et al., 2022a).

Research has demonstrated that introducing AM fungi into the soil can benefit plant growth, particularly in conditions where phosphorus is limited, by improving P uptake (Spoorthi et al., 2022; Bagyaraj et al., 2022a). AM fungi have also been found to enhance plant tolerance or resistance to abiotic and biotic stressors such as drought, salinity and metal toxicity, as well as to root pathogens (Begum et al., 2019; Weng et al., 2022; Malviya et al., 2023). The fungi contribute to the formation of stable soil aggregates, promote a macroporous soil structure, facilitate water and air penetration, and reduce erosion. Beyond facilitating P uptake, AM fungi also aid in the availability of other essential elements, such as Zn, Cu, K, Al, Mn and Fe (Bagyaraj, 2011; Sheeba et al., 2022). They are also known to impact plant hormone levels and enhance water relationships in plants. However, it remains unclear whether the observed benefits are directly attributed to the fungus or indirectly result from improved P nutrition that affects host physiology.

Research has shown that mycorrhizal plants experience numerous physiological and anatomical changes (Bahadur et al., 2019). Additionally, the presence of AM fungi in plant roots can alter root exudations and impact rhizosphere microorganisms, further benefiting plant growth (Nunes, 2018). These findings demonstrate the numerous ways in which AM fungi can benefit host plants and improve overall plant growth.

Isolation and Maintenance

To determine if a root system is colonized by AM fungi, visual inspection is not sufficient as there are no noticeable morphological differences. However, in some plant species like onion, maize and clover, root colour may be yellow but disappears quickly on exposure to light. Therefore, the mycorrhizal status of a root system can only be confirmed through microscopic observation after preparing and staining the roots with trypan blue or other suitable stains (Smith and Dickson 1997).

AM fungal spores can be isolated from soil by using the wet sieving and decantation method, as described by Bagyaraj and Sturmer (2008) and Trejo-Aguilar and Banuelos (2020). The single AM fungal spore isolation method involves surface sterilizing single spores obtained from the soil sample with a solution containing chloramine T and streptomycin sulfate, followed by picking individual spores with a fine capillary pipette and placing them in a funnel filled with sterile sand and seeding with a suitable host. The seedlings are then transferred to a sterilized sand: soil mixture in a pot and maintained in a greenhouse as a "pot culture." Once a pot culture of a single species is established, it can be used as an inoculum to multiply the fungus for future experiments. This method allows for the isolation and cultivation of single AM fungal species for further research (Agnihotri et al., 2022).

The trap culture technique is an effective alternative method for isolating AM fungi. This method involves enriching the mycorrhizal propagules collected from the root zone soil with the suitable host under greenhouse conditions which yields numerous healthy spores of colonizing fungi that can be utilized for identification, establishing monospecific cultures and also for various experiments. The trap culture technique provides a controlled environment that encourages sporulation and optimizes the collection of healthy spores; unlike spores collected directly from field soil, which may have viability, appearance and representativeness issues. Factors such as root pigments, soil chemistry, temperature, moisture, and microbial activity may affect spores' structural characteristics and viability. Furthermore, spores only represent arbuscular fungi with adequate activity and biomass to trigger sporulation. By providing a controlled environment, the trap culture technique can overcome these challenges (Dsouza, 2019; Tenzin et al., 2022).

Mycorrhizal Dependency

Mycorrhizal dependency refers to the extent to which a plant species depends on the symbiotic relationship it forms with mycorrhizal fungi to survive and flourish. Mycorrhizal fungi reside in the soil and establish associations with plant roots, providing the plant with water and nutrients while receiving carbohydrates produced by the plant through photosynthesis. Some plant species display high mycorrhizal dependency and reliant on these fungi for nutrient uptake, while others have a lower mycorrhizal dependency and can exist without this symbiotic relationship. For instance, seedlings of *Liquidambar stryraciflua* are obligately mycorrhizal, whereas vegetatively propagated cuttings of *Liriodendeon tulipifera* mostly remain dormant unless made mycorrhizal (Kormanik et al., 1980).

Gerdemann's (1975) definition of relative mycorrhizal dependency (RMD) explains the extent to which a plant relies on mycorrhizal fungi to produce maximum growth or yield at a particular level of soil fertility. Menge et al., (1979) proposed a formula that used an inoculant AM fungus in sterilized soil to calculate the mycorrhizal dependency of crop plants. However, the limitation of this formula is that the RMD values can exceed 100%, making it challenging to categorize crop plants. Plenchette et al. (1983) introduced another formula to calculate the relative field mycorrhizal dependency (RFMD) of crop plants under field conditions by comparing plants in fumigated and non-fumigated soils. This formula measures the growth increase due to native endophytes and provides practical insight into the response of crop plants to native mycorrhiza before applying inoculation to crops. Despite the presence of endophytes in unsterile soil conditions, the introduction of mycorrhizal fungi has been reported to improve plant growth. Hence, Bagyaraj et al. (1988) proposed another formula to assess the growth improvement brought about by inoculation with a mycorrhizal fungus in unsterile soil with indigenous AM fungi, called the mycorrhizal inoculation effect (MIE).

The degree of mycorrhizal dependency in plants can fluctuate depending on a range of factors, including soil conditions, plant species and environmental stressors. Mycorrhizal dependency is an essential factor to consider in agriculture and land management since it can significantly impact plant growth, productivity, soil health, and nutrient cycling. Having a comprehensive understanding of the mycorrhizal dependency of various plant species can assist in making decisions regarding planting and management practices that facilitate healthy plant-fungi associations. Additionally, this knowledge is vital for developing sustainable agricultural practices that prioritize the conservation and enhancement of soil health (Tawaraya, 2003; Ortas, 2019).

Need for AM Fungal Inoculation

After recognizing the importance of AM symbiosis, a decision needs to be made on whether to use the native population of AM fungi as a starting material or to supplement it with external inoculation. If the native fungal population is sufficient, then its efficacy in terms of improving plant growth should be assessed. Spore counts are generally used to assess the native fungal population, but this method does not provide a clear indication of the population of AM fungi as sporulation depends on host and environmental factors (Bagyaraj, 2011). Another option is to collect young, fresh roots of native plants growing in the soil and microscopically examine them for AM colonization after staining, but this may not always be possible. The extent of root colonization by a particular AM fungal species can vary depending on the host, fungal, and environmental factors, and the absence of colonization at a particular time does not necessarily indicate that the soil is devoid of AM fungi. The best way to determine the effectiveness of native AM fungi would be to grow the intended crop in the soil to which it would be transplanted and evaluate the benefits for plant growth derived from the native AM fungi.

Some of these fungi are reported to cause no growth improvement despite maximum root colonization (Lovato et al., 1992). Hence, the selection of an efficient fungus is crucial for obtaining maximum benefits from the plant-fungal association. Host preference and variation in the extent of root colonization and benefits conferred to host plants by different AM fungal species and isolates make it necessary to screen and select an efficient fungus for each plant species (Bagyaraj and Kehri, 2012). In some cases, the native AM fungal inoculum or species may not be efficient in enhancing the growth and nutrition of host plants to a great extent as compared to the introduced AM inoculum. Therefore, it may be necessary to inoculate the substrate with a selected efficient fungus or combination of strains to obtain maximum benefits (Bagyaraj and Kehri, 2012; Ranadev et al., 2022).

Selection of Efficient AM Fungi

The selection of particular AM fungi for inoculation into agricultural soils should be based on their ability to enhance nutrient uptake by plants and persist in soils. Effectiveness, extent of colonization and survival in the rhizosphere and soil are important characteristics required by inoculant fungi, similar to the requirements for successful *Rhizobium* inoculum strains for legumes. However, the selection process for AM fungi is not always clearly explained in research studies. A thorough discussion on the selection of AM fungi for use in agriculture and forestry has been published earlier (Bagyaraj and Kehri 2012; Bagyaraj et al., 2022a).

The screening of AM fungi for efficient symbiotic response using a test host followed by pot culture, microplot, and field trials can help in selecting fungi for pre-inoculation of transplanted crops raised in unsterile soil. It is important to add the same number of infective propagules of different fungi based on most probable number counts, and to include AM fungi isolated from the root zone of the test host in the selection process. This procedure has been successful in selecting efficient fungi for many economically important agriculture crops like soybean, maize, cowpea; horticulture crops like tomato, chilly, capsicum and tree species such as *Leucaena leucocephala, Tamarindus indica, Acacia nilotica, A. auriculiformis, Calliandra calothyrsus, Casuarina equisetifolia, Azadirachta indica, Dalbergia sissoo, D. latifolia* and *Tectona grandis* (Bagyaraj and Kehri, 2012; Muleta and Diriba, 2017; Raghu et al., 2021; Emmanuel et al., 2020; Bagyaraj et al., 2022a; Wang et al., 2022).

It is important to note that the selection of efficient AM fungi is not a one-time process, as the effectiveness of a fungus can vary depending on environmental factors and the host plant species. Therefore, it is necessary to regularly evaluate the performance of selected fungi in different environments and with different host plants. Additionally, the use of multiple strains or species of AM fungi in inoculation may provide synergistic benefits to the host plant, as each fungus may have different abilities to mobilize and transfer nutrients. Overall, the selection of efficient AM fungi is a critical aspect of mycorrhizal technology that can have a significant impact on agricultural and forestry practices (https://invam.ku.edu/ accessed as on May 06, 2023).

P Response Curves

The P response curve of mycorrhiza refers to the relationship between the level of P in the soil and the response to mycorrhizal fungi. The P response curve of mycorrhiza typically shows an initial rapid increase in mycorrhizal colonization and P uptake by the plant as the soil P level increases from low to moderate. However, as the soil P level continues to increase, the benefits of mycorrhizal fungi to the plant may decrease, or even become negative, as the plant may rely more on direct uptake of P from the soil. The shape of the P response curve can vary depending on the specific mycorrhizal species and the plant species involved, as well as other environmental factors such as soil type and pH. Understanding the P response curve of mycorrhiza can be important for optimizing P fertilizer applications in agriculture and managing soil P levels for plant growth (Berger and Gutjahr, 2021).

Plant Growth Response to Inoculation of AM Fungi and Mechanisms Involved in Plant Growth Promotions

Previous experiments conducted in sterilized soil have shown that inoculating with AM fungi can improve plant growth. It was once believed that since most natural soils already contain AM fungi, plants would not respond to mycorrhizal inoculation in unsterilized soils. However, subsequent investigations have indicated that even in unsterilized soils, plants do respond to efficient AM fungal strains. It has now been conclusively proven that AM fungi significantly enhance plant growth by increasing the uptake of diffusion-limited nutrients such as P, Zn, Cu and also other nutrients from the soil. Other beneficial effects of AM fungi include their role in the biological control of root pathogens, hormone production, increased ability to withstand abiotic stress, and synergistic interaction with nitrogen fixers, phosphorus solubilizers and PGPR (Kuila and Ghosh, 2022).

Tropical soils generally have low inherent fertility and are often deficient in P, making the role of AM fungi in improving plant growth more significant in these regions than in temperate soils. Additionally, phosphorus in tropical soils can become fixed over time, further reducing its availability to crops. In acidic soils, phosphorus can become fixed as iron and aluminium phosphates, while in neutral soils, it becomes fixed as calcium phosphates. The continuous use of P fertilizers can result in the accumulation of fixed P in the soil, creating large reserves over time (Berruti et al., 2016; Bagyaraj et al., 2022b). A small proportion (less than 10%) of soil P is utilized in the plant-animal cycle. Studies involving ^{32}P-labeled P have shown that AM fungi cannot solubilize unavailable inorganic P sources, but they can extract additional phosphate from the soil solution's labile pool (Qin et al., 2022). Hyphae of AM fungi extend beyond the root system, allowing for the exploration of spatially unavailable nutrients and improving plant P nutrition (Bagyaraj et al., 2022b). As plant roots absorb P from the soil solution faster than it moves in soil solution by diffusion, a P depletion zone forms around the root. AM fungi play a crucial role in this zone by extending their hyphae beyond it, scavenging a large soil volume, and providing phosphorus to plants in exchange for carbohydrates from their host plant (Bagyaraj and Ashwin, 2017). In general, AM fungi are particularly beneficial in soils with low to moderate fertility, particularly in P limiting concentrations (Bagyaraj et al., 2015), as it promotes plant growth.

AM fungi have been shown to have a significant impact on the physiology and anatomy of plants, leading to increased rates of respiration, photosynthesis and production of essential compounds such as sugars, amino acids and RNA. The increased chloroplasts, mitochondria, xylem vessels and motor cells were observed in AM fungal inoculated plants (Nunes et al., 2018; Bahadur et al., 2019). These fungi can also alter root exudations and affect the microbial community in the rhizosphere, further promoting plant growth. Inoculation with AM fungi can also maintain high populations of beneficial soil organisms, including *Azotobacter, Azospirillum* and phosphate-solubilizing bacteria and other PGPR which can have synergistic effects on plant growth. Additionally, AM fungi have been found to have the ability to deter or reduce the severity of disease caused by soil-borne pathogens, highlighting the multiple benefits that these fungi offer to host plants for overall improved growth (Berruti et al., 2016; Malviya et al., 2023).

Research has shown that AM fungi can upregulate membrane transporter genes, as well as genes involved in phosphate uptake, such as GiPT, GmosPT and GvPT, which have been identified in the extraradical mycelium (ERM) of *G. intraradices, G. mosseae* and *Glomus versiforme* respectively (Benedetto et al., 2005; Giovannini et al., 2022). Additionally, Proton-ATPases (H^+-ATPases) responsible for Pi uptake across the plasma membrane of ERM have been identified in *G. mosseae* (Seemann et al. 2022; Rui et al., 2022). Studies have also reported an increase in total amino acid concentrations in leaves of tea accompanied by the upregulation of genes involved in amino acid synthesis, such as glutamate dehydrogenase (CsGDH), glutamate synthase

(CsGOGAT), and glutamine synthetase (CsGS) (Li et al., 2022). An analysis of 6124 differentially expressed transcripts (DETs) using Oxford Nanopore Technologies (ONT) MinION revealed that 391 DETs were specifically regulated by AM fungus under salinity stress conditions. This offers molecular evidence of salt tolerance provided by AM fungi. These genes primarily participate in improving the internal environment in plant cells, nitrogen metabolic-related processes, possible photoprotection mechanisms and in regulating root ROS-scavenging capacity (Zhang et al., 2021).

Application Methods and Time of Application

Standardization of AM fungal inoculum dose and application time is crucial to achieve maximum benefits from the AM fungal-host association. A higher density of infective propagules in the inoculum can ensure quicker and adequate colonization of the host root. A density of about 150-200 infective propagules per gram of substrate (sand: soil 1:1) is recommended for good colonization and establishment of the host, although this may vary depending on factors such as soil pH, nutrient status, light intensity, and temperature. For horticultural and forest tree species with larger planting distances, a higher inoculum density is preferred, while in agricultural crops with higher plant density, lower inoculum may be sufficient.

It has been well demonstrated that the earlier the inoculation is applied, the greater the benefits to the host. Earlier studies have shown that inoculating AM fungi at the time of sowing/planting results in better root colonization leading to better establishment of the host. In experiments conducted with micropropagated *Ficus benzamina*, it was found that the optimal time for inoculation is just before planting out hardened plantlets (Shashikala et al., 1999). Methods of applying AM fungi generally include hand placement, placing below the seed or seedling in transplanted crops. The importance of the method of inoculum application arises from the need to initiate colonization in the early stages of plant growth. Therefore, it is essential to place the inoculum close to the seed material to ensure that the roots come into contact with the fungal material as soon as they emerge. Based on an experiment, it was concluded that the best method of inoculum placement for seedlings raised in polybags was placing the AM inoculum 8 cm below the surface of the soil (7 cm below the seed), at a point or as a layer. For seedlings raised on nursery beds, placing the inoculum 2 cm below the seed was found to be the most effective approach (Bagyaraj, 2011).

It is important to note that the effectiveness of AM fungal inoculation may also depend on the type of plant species, soil type and management practices. Therefore, it is recommended to conduct trials to optimize the dosage and application of AM fungal inoculum for each specific crop and soil type. Additionally, the source of inoculum should also be considered, as not all AM fungal species are suitable for all plants and soil types (Thilagar and Bagyaraj, 2015; Ranadev et al., 2022). The quality of inoculum should also be checked, as it can be contaminated with other harmful microorganisms or may have low viability. Overall, proper management of AM fungal inoculation can lead to significant improvements in plant growth and health. Furthermore, it is important to ensure that the inoculum is viable and has the potential to colonize plant roots rapidly. The inoculum should be stored under proper conditions, such as room temperature and adequate moisture to maintain its viability. Before application, it is necessary to test the inoculum for its quality and infectivity and ensure that it contains viable propagules of the desired AM fungal species. Quality control measures, such as periodic monitoring of the inoculum, should also be implemented to ensure its effectiveness (Agnihotri et al., 2022).

In summary, the optimal dose, time and method of application of AM fungi are important factors to consider in order to maximize the benefits of the AM-host association. The inoculum should be of high quality and infectivity and should be applied close to the seed material to initiate colonization in the early stages of plant growth.

Appropriate Technology for Nursery Raised Crops

Inoculation of nursery beds with suitably selected strains of AM inoculum is a simple and effective method for growers. The inoculum can be incorporated into the nursery beds or containers at the appropriate rate by hand and the resulting seedlings will be colonized by the introduced fungus before being transplanted to the field. In developed countries, seedlings are typically raised in fumigated soil or potting mix. Inoculating this substrate with selected AM fungi for the relevant crop or forest tree species is an appropriate and straightforward technology. Inoculating unsterilized potting mix with appropriate fungi has been shown to result in vigorously growing seedlings that perform well when transplanted to the field. Inoculating unsterile nursery beds with efficient AM fungi has been found to enhance the growth of several horticultural crops, forest tree species and medicinal plants, as well as plantation crops (Lakshmipathy et al., 2000; Tilak, 2010; Bagyaraj, 2011; Anuroopa et al., 2017 Bagyaraj et al., 2022a).

Production of Seedlings Inoculated with AM Fungi

Forest plants that are propagated by seedlings can benefit from inoculation with AM fungi, as it requires a small amount of inoculum and is easy to handle. Growing the seedlings in semi-controlled conditions can help in the establishment of inoculated AM fungi. Additionally, using subsoil and organic material in the substrate composition, which have few or no AM fungal propagules, reduces competition with native AM fungi. In many cases, the substrate is disinfected to prevent pathogens and weeds which alsokills the native AM fungi. Inoculating seedlings is particularly important for plants with high mycorrhizal dependency or for those planted in areas with few AM fungal propagules, such as degraded areas or those with a long history of cultivating a non-AM species (Bagyaraj and Ashwin, 2017).

Further, inoculation of seedlings with AM fungi has multiple benefits beyond the improvement of plant growth and nutrition. It can also increase uniformity of seedlings, enhance tolerance and protection against soil pathogens and reduce the time and cost of nursery production. However, the success of seedling inoculation depends on several technical aspects, including the choice of inoculum and handling of inoculated seedlings, as well as the adaptation of the technology to local production conditions. Ultimately, the aim of seedling inoculation should be the production of high-quality seedlings that can provide post-transplantation benefits, rather than simply the benefit of inoculation during the seedling production phase. The production of mycorrhizal seedlings can be particularly beneficial for reforestation, land restoration and sustainable agriculture (Bagyaraj and Ashwin, 2017; Raghu et al., 2021).

Use of AM Fungi in Rooting of Cuttings, Air Layering and Overcoming Transplant Shock

Propagation of some forest and horticulture plants involves the use of cuttings and in such cases, the rooting of the cuttings is crucial for successful growth. Studies have shown that inoculating cuttings with AM fungi can enhance rooting in plants (Omar, 2007; Caruso et al., 2021), while reports have also shown that air-layered tamarind and cashew plants display increased rooting (Bagyaraj and Mallesha, 2000). These plants not only tolerate transplant shock better but also establish more effectively when planted in field sites. Additionally, earlier research reported that AM-inoculated avocado and cashew plants demonstrated better tolerance to transplant shock and survival compared to uninoculated plants (Lakshmipathy et al., 2000). It has been reported that AM-inoculated seedlings of *Eucalyptus globulus*, showed greater survival rates and higher growth performance in the field compared to uninoculated seedlings (Azcon et al., 1996). Inoculation with AM fungi can also reduce the need for fertilizer inputs, as the plants are able to take up more nutrients from the soil through their mycorrhizal associations. Overall, the use of AM fungi in nursery and propagation practices can lead to improved plant growth, establishment and survival as well as reduced production costs and environmental impacts.

Inoculation of Directly Field Sown Crops

There is mounting evidence that AM fungi can significantly enhance crop productivity in the field and are crucial to the way crops respond to fertilisation regimes. Research has shown that the application of AM fungi can reduce the use of phosphatic chemical fertilizers by up to 50 percent (Thilagar et al., 2016). AM fungi offer numerous benefits to plants, including improved nutrient uptake, increased tolerance to environmental stresses such as drought and salinity, and enhanced disease resistance. Additionally, AM fungi can improve soil physicochemical properties by producing soil proteins known as glomalin (Fall et al., 2022). Given these benefits, there is growing interest in using AM fungi as a sustainable alternative to chemical fertilizers and pesticides in agriculture. Overall, AM fungi play a vital role in soil ecosystems and are a valuable resource for promoting plant growth and productivity (Ranadev et al., 2022; Agnihotri et al., 2022).

Improved plant growth because of AM fungal inoculation has also been reported in crops like fodder maize, sorghum, alfalfa, *Hedysarum coronarium* and cowpea (Ranadev et al., 2022). The application of AM fungi under field conditions has been shown to enhance the yield of several crops like 25-30% in tomato (Bona et al., 2017), 14-17% in chilli (Thilagar et al., 2016), 17-20% in soybean (Ashwin et al., 2023), 20-22% in garlic (Borde et al., 2009), 13-24% in cowpea (Haro et al., 2020) and 16-20% in potato (Douds et al., 2007). The field application of AM fungi not only increases the growth and yield but also increases the effectiveness of fertilizer application and reduces the fertilizer dosage (Cely et al., 2016).

Direct inoculation of AM fungi was earlier limited to small-scale, high-value crops and hence there is limited information on large-scale field inoculations, since these fungi are obligate symbionts, which hinders the production of inoculum. However, this situation is rapidly changing with the development of new approaches and methods to improve both inoculum production and application. Direct application of AM fungi to field-sown crops can be achieved through seed treatments or soil inoculations and the choice of application method will depend on the crop, soil type, and environmental conditions. Overall, the application of AM fungi to directly field-sown crops can be a cost-effective and sustainable way to enhance crop productivity and reduce environmental impacts (Estaún et al., 2002; Cely et al., 2016).

Interaction of AM Fungi with Other Beneficial Soil Microorganisms

AM fungi can interact with a wide range of other beneficial soil microorganisms, such as nitrogen fixers, phosphate solubilizers and PGPR forming complex microbial communities that can benefit plant growth and ecosystem health. Studies on the interaction of AM fungi with other beneficial microorganisms suggest that these are synergistic, resulting in significant benefits for plant growth (Dames and Ridsdale, 2012; Bagyaraj et al., 2022a). These interactions can have important implications for plant growth and health, as well as for the overall functioning of soil ecosystems. Recent studies indicate that inoculating crops with a microbial consortia composed of an efficient AM fungus, nitrogen fixer, P solubilizer and PGPR, all carefully selected and screened for a particular crop plant or forestry species, is more beneficial than using only AM fungus in terms of improving growth, biomass and yield (Raghu et al., 2021; Bagyaraj et al., 2022b).

One of the most important interactions of AM fungi is with nitrogen-fixing bacteria. AM fungi can enhance the growth and activity of these bacteria, which can in turn provide plants with an additional source of nitrogen. This can be particularly important in nitrogen-poor soils, where plants may struggle to obtain enough nitrogen for their growth. Synergistic interaction between root nodule bacteria (*Rhizobium* spp.) and AM fungi in leguminous crop plants is well documented (Gupta et al., 2007). Inoculation with AM fungi increased nodulation and nitrogen fixation by rhizobia and inoculation with rhizobia increased colonization and P uptake by AM fungi. It has been reported that in agriculture crops such as soybean, maize, cowpea; horticulture crops such as tomato, chilly, capsicum and forest tree species such as *Acacia, Robinia, Prosopis* and *Leucaena* that dual inoculation with AM fungi and rhizobia increased nodulation, N and P uptake and dry plant biomass (Muleta and Diriba, 2017; Emmanuel et al., 2020; Ashwin et al., 2022; Wang et al., 2022). Previous studies have shown that the dual inoculation of *Funneliformis mosseae* + *Bacillus sonorensis* at the time of transplanting not only enhances plant growth and yield, but also reduces the use of NPK fertilizers by 50% under field conditions, which improves soil health and reduces environmental pollution (Thilagar et al., 2015; Jyothi et al., 2018).

AM fungi can also interact with other types of fungi in the soil, such as saprotrophic fungi and other mycorrhizal fungi. These interactions can be both positive and negative. For example, AM fungi can stimulate the growth of saprotrophic fungi, which can help to break down organic matter in the soil and release nutrients for plant uptake. However, they can also compete with other mycorrhizal fungi for resources, which can limit their effectiveness. The various synergistic interactions of AM fungi with other beneficial soil organisms like P-solubilizers, PGPR, and biocontrol agents are well documented (Wang et al., 2022)

In addition to these interactions, AM fungi can also play an important role in shaping the overall structure and function of soil microbial communities. For example, they can modify the pH of the soil, which can in turn affect the growth of other microorganisms. They can also produce various compounds, such as glomalin, that can help to stabilize soil aggregates and protect soil from erosion (Yang et al., 2017). Overall, the interactions of AM fungi with other beneficial soil microorganisms are complex and multifaceted. However, these interactions are crucial for maintaining healthy soil ecosystems and supporting the growth and health of plants.

AM Fungi for Biocontrol of Plant Pathogens

There are multiple reports on the interactions between AM fungi and root pathogens suggesting that the severity of diseases can be reduced or mitigated by AM fungi. Such consistent reduction in disease symptoms has been observed for fungal, bacterial and nematode pathogens. Mechanisms of suppression may include morphological, physiological and biological alterations in the host, such as the thickening of cell walls through lignification and production of other polysaccharides, which prevent the penetration and growth of pathogens like *Fusarium oxysporum* and *Phoma terrestris* (Frac et al., 2018). Furthermore, mycorrhizal plants have a higher concentration of ortho-dihydroxy phenols compared to non-mycorrhizal plants, which has been found to be inhibitory to the root rot pathogen *Sclerotium rolfsii*.

AM fungal colonization can activate specific plant defence mechanisms, such as phytoalexins, enzymes of the phenylpropanoid pathway, chitinases, peroxidases and pathogenesis-related (PR) proteins, thereby providing a protective capacity against pathogens. Mycorrhizal plants also have a higher population of microorganisms in the rhizosphere, making it difficult for the pathogen to compete and gain access to the root. Additionally, the mycorrhizosphere supports a higher population of antagonists and siderophore producers, suggesting that biological control of root pathogens with AM fungi is a promising possibility (Aljawasim et al., 2020; Muthukumar et al., 2022).

Mass Production of AM Fungi

Supply of adequate good quality AM fungal inoculum is crucial for successful sustainable agriculture. The availability and quality of inoculum play an important role in improving plant and soil health. AM fungi are currently being produced through various methods *viz.*, substrate-based pot cultures (Coelho et al., 2014), on-farm (Douds et al., 2005,), aeroponics (Hung and Sylvia 1988; Agnihotri et al., 2022), nutrient film technique (NFT) and root organ culture (Fortin et al., 2002, Ijdo et al., 2011). However, on a commercial scale root organ culture technique (*in vitro*) and substrate-based pot cultures (*in vivo*) are being used. Inoculums produced through these modes are commercially available to farmers by many firms across the globe (Marleen et al., 2011).

In-Vivo Method of Mass Production or Substrate Based Inoculum

The traditional "Pot Culture" technique is used to produce substrate-based inoculum, which contains all AM fungal structures in a highly infectious state (Trejo-Aguilar and Banuelos, 2020). Choice of host plant in the propagation of AM fungi in pot cultures is an important decision. The obligate nature of the symbiosis prevents separation of host variables from those of soil and ambient environments as they impact interactively on growth and sporulation of colonizing fungi. Based on the results of several experiments Rhodes grass (*Chloris gayana*) and Sudangrass (Sorghum × Drummondii) were found to be the best host, while vermiculite: perlite:soilrite (3:1:1 v/v) was the best substrate (Selvakumar et al., 2016). Calcium ammonium nitrate and rock phosphate were identified as the best N and P sources for mass-producing *Glomus fasciculatum*. Additionally, using a modified Ruakura nutrient solution added once in 8 days and harvesting pot cultures in 75 days resulted in high-quality inoculum with the maximum number of infective propagules (Bagyaraj et al., 2022a).

In-Vitro Method of Mass Production or Root Organ Culture

The *In vitro* mass production of mycorrhizal fungi involves growing the fungi in a sterile laboratory environment. The process begins by establishing a culture of the desired mycorrhizal fungi, typically using spores or mycelium obtained from a natural source in the hairy roots. The fungi infected hairy roots are then grown in a nutrient-rich medium, such as modified Strullu-Romand medium, under controlled conditions of temperature, light, and humidity. As the fungi grow, they produce spores, which can be harvested and purified for use as inoculum. *In vitro* mass production of mycorrhizal fungi has several advantages, including the ability to produce a large number of spores in a relatively short time and the ability to maintain pure cultures of specific fungal strains (Kokkoris and Hart, 2019). This method is commonly used in research and commercial applications, such as the production of mycorrhizal inoculants for the use in agriculture, horticulture and forestry. Some of the disadvantages associated with *in vitro* mass production are lesser shelf life, reduced genetic diversity, morphological and functional alterations *viz.* reduced propagule size, lower infectivity, and the difficulty in mass producing all AM fungi by this method. The commercial production of AM fungi *via* Transformed root cultures (TRC) or root organ culture represents strong selection pressure on fungi and represents a form of domestication, through changes to nutrient limitations and reduced host variation. Such selection pressure may lead to reduced genetic diversity and mutualistic quality (Kokkoris and Hart, 2019; Agnihotri et al., 2022).

Quality Assessment of AM Fungal Inocula

The mass production of arbuscular mycorrhizal (AM) fungi has recently become a large biofertilizer industry. Because these fungi have an obligate symbiotic nature, they are primarily propagated on living roots in substrate-based pot cultures and to a limited extent through Ri T-DNA in *in vitro* or root organ culture systems. The determination of AM fungal inoculum quality is a critical aspect of inoculum production. The quality control of AM fungal inoculum viz., enumeration, identification and purity estimation is important for their large scale production (Crossay et al., 2017). The identification of compatible strains should be a prerequisite before the field application of AM fungi. Furthermore, the inoculum quality evaluation should encompass microscopic, biochemical and molecular methods. Other critical aspects that govern the quality of AM fungal inoculum and the success of inoculation in the field include i) soil nutrient profile and fertilizer dose; ii) efficiency and abundance of native AM fungal population; iii) the extent of host dependence on AM fungi/mycotrophic capacity of the plant and iv) inoculum application rate and viability (Tarbell and Koske, 2007). The measures highlighted above will contribute to the large-scale production and successful application of high-quality AM fungal inoculum in agriculture.

Conventional microscopic methods, such as spore numbers and infective propagule numbers/g, are commonly used to assess the quality of the inoculum. However, newer biochemical methods (phospholipid fatty acid profiling) and molecular methods (qPCR) have been attempted, but they are still in the research stage. The most frequently encountered problems with these methods include a lack of primers, signature fatty acids representing more than one microbial species, and the presence of different infectious propagules by species within root segments (Agnihotri et al., 2022). *In vitro* production ensures the purity of AM fungal isolates, but it remains technically rigorous and is not currently feasible for all AM fungal species. Inoculum produced in this manner often consists of spores of one AM fungal species. Substrate-based production is the method of choice for mass-producing AM fungi in developing countries because of its easy method of production, contains all infective propagules, may consist of several AM fungal species and has a long shelf-life at room temperatures (Douds et al., 2005; Agnihotri et al., 2022).

Considering the pros and cons of the different methods available for testing the quality of commercially produced AM fungal inoculum, determining the infective propagule numbers/g by the MPN method with 10-

fold dilution appears to best reflect the quality of the inoculum (Agnihotri et al., 2022). A combination of these methods is often used to provide a comprehensive assessment of the quality of AM inocula. In India, according to the fertilizer control order (FCO) specifications, AM fungal biofertilizers should contain a minimum 10 viable spores and 1200 infective propagules (IP) per gram of inoculum [Fertiliser (Inorganic, Organic or Mixed) Control Fifth Amendment Order, 2021]. These requirements differ in different countries.

Conclusion

For several decades, chemical fertilization has been used in agriculture to improve crop yields, resulting in environmental contamination and soil health degradation. As a result, sustainable agriculture has grown in importance, relying on fewer chemicals and more biological alternatives such as compost, biofertilizers, and biopesticides. Among these biofertilizers, AM fungi play an important role in improving plant growth by increasing the uptake of diffusion-limited nutrients (Chifetete and Dames, 2020), interacting synergistically with beneficial soil microorganisms, producing plant growth-promoting substances, and assisting plants in coping with water, salinity, heavy metal stress, and root pathogens. However, the low quality of commercial inoculum is a major impediment to the use of AM fungi. *In vitro* production of AM fungi ensures the purity of isolates but remains technically rigorous. Mass production of AM fungi by *in vivo* method appears to be the best at present. Many farmers are unaware of how to utilize mycorrhizal inoculum, emphasizing the importance of education. This could be achieved through demonstration trials on farmers' fields and media publicity such as TV programs, local public articles and seminars at growers' meetings. The use of AM fungi as biofertilizers will not only improve plant growth and productivity but also minimize the use of fertilizers and pesticides, reducing environmental pollution.

Future Directions of Work

Given below are the future directions of work in AM fungi which will help in utilizing it in agriculture and allied areas.

- The specific surveys of natural mycorrhizal associates of diverse crops in different climatic and soil zones and in natural ecosystems of the world. Along with the traditional spore-based taxonomic methods, molecular metagenomic approaches to identify the species (Alrajhei et al., 2022).
- The composition of the AM fungal community is strongly affected by agricultural management. A clear negative correlation between land use intensity and the diversity of AM fungal communities has been demonstrated on a global scale (Wright et al., 2005). Hence the research should focus on finding sustainable management strategies to facilitate the better utilization of AM fungi in agriculture.
- The investigations on the AM fungal hyphosphere microbiome, associated with the extraradical hyphae of AM fungi, which is home to a diverse range of microbes. Understanding the functional roles of hyphosphere microbiome in soil carbon and nutrient cycling, plant nutrition and health, and soil food web dynamics, including the interactions and mechanisms facilitating microbial co-occurrence and co-operation (Faghihinia et al., 2023).
- Exploring the potential of AM fungi for alleviation of biotic and abiotic stresses, phytoremediation and ecological restoration of degraded soils (Ren et al., 2019).
- Studies conducted in 1980s on cowpea genotypes indicated that AM fungal colonization is not only host-dependent but also a heritable trait and can be used in plant breeding programmes (Mercy et al., 1990). In 2001 it was reported that older wheat varieties with a lower uptake capacity of phosphate are more responsive to AM fungi than newer varieties with a higher endogenous uptake capacity (Zhu et al., 2001). In the case of maize varieties, however, modern hybrids are colonized more extensively than old landraces and genotypes adapted to low phosphate environments were less responsive to AM fungi than genotypes adapted to high phosphate environments (Wright et al., 2005). Hence, use of genotypes from high phosphate environments with a high responsiveness for AM fungi can be used for future breeding approaches.
- AM fungi have been found to enhance carbon sequestration in soil, which could help mitigate the effects of climate change (Maússe Sitoe and Dames, 2022). Future research could explore the potential of AM fungi to sequester carbon in different ecosystems and under different climate scenarios.

- In recent years application of beneficial microbial consortia is gaining more attention because of wider benefits than single species application. Hence there is an urgent need to develop the mycorrhizal consortia or microbial consortia with AM fungi for various crop plants keeping the factors in mind viz. soil type, crop and agro-climatic zones.
- The development of efficient and scalable methods for the production and application of AM fungal inoculum (Agnihotri et al., 2022).
- Identifying the genes and transcription factors of both AM fungi and host plants that enable the AM fungi to react to diverse environmental stresses has become an essential task for researchers in this field for the future (Sun et al., 2022). Such investigations including CRISPR-Cas technology would enable the development of mycorrhizal partnerships with a maximum positive effect on host plants (Geetha and Dathar, 2022).
- Improving our understanding of the economic and societal impacts of AM fungi including the development of sustainable and socially responsible business models for the production and distribution of AM fungal inoculum.

References

Abbott, L. K., and Robson, A. D. (1984). Infectivity and effectiveness of five endomycorrhizal fungi: Competition with indigenous fungi in field soils. *Aust J Bot*, 32(5), 621-30.

Agnihotri, R., Sharma, M. P., Bucking, H., Dames, J. F., and Bagyaraj, D. J. (2022). Methods for assessing the quality of AM fungal bio-fertilizer: Retrospect and future directions. *World J Microbiol Biotechnol*, 38(6), 1-14.

Ahmad, M., Summuna, B., Gupta, S., and Sheikh, P. A. (2020). Mycorrhizal Association: An Important Tool for the Management of Root Diseases. *Int J Curr Microbiol App Sci*, 9(9), 2020.

Aljawasim, B., Khaeim, H. M., and Manshood, A. M. (2020). Assessment of arbuscular mycorrhizal fungi (*Glomus* spp.) as potential biocontrol agents against damping-off disease *Rhizoctonia solani* on cucumber. *J Crop Prot*, 9(1), 141-147.

Alrajhei, K., Saleh, I., and Abu-Dieyeh, M. H. (2022). Biodiversity of arbuscular mycorrhizal fungi in plant roots and rhizosphere soil from different arid land environment of Qatar. *Plant Direct*, 6(1), 361-369.

Anuroopa, N., Bagyaraj, D. J., Abhishek, B., and Prakasa, R. (2017). Inoculation with Selected Microbial Consortia Not Only Enhances Growth and Yield of *Withania somnifera* but also Reduces Fertilizer Application by 25% Under Field Conditions. *Proc Ind Natl Sci Acad*, 83, 957-971.

Ashwin, R., Bagyaraj, D. J., and Mohan Raju, B. (2022). Dual inoculation with rhizobia *Bradyrhizobium liaoningense* and arbuscular mycorrhizal fungus *Ambispora leptoticha* improves drought stress tolerance and productivity in soybean cultivars MAUS 2 and DSR 12 under microplot conditions. *Biologia*, https://doi.org/10.1007/s11756-022-01196-3.

Ashwin, R., Bagyaraj, D. J., and Mohan Raju, B. (2023). Ameliorating the drought stress tolerance of a susceptible soybean cultivar, MAUS 2 through dual inoculation with selected rhizobia and AM fungus. *Fungal Biol Biotech*, https://doi.org/10.1186/s40694-023-00157-y.

Azcon Aguilar, C., and Barea, J. M. (1996). Arbuscular mycorrhizas and biological control of soil borne plant pathogens – an overview of the mechanisms involved. *Mycorrhiza*, 6, 220-223.

Bagyaraj, D. J., and Ashwin, R. (2017). Can mycorrhizal fungi influence plant diversity and production in an ecosystem. In: Microbes for restoration of degraded ecosystems. *NIPA*, 1-7.

Bagyaraj, D. J., and Kehri, H. K. (2012). *Microbial Diversity and Functions*. New India Pub Agency, 641-667.

Bagyaraj, D. J., Sridhar, K. R., and Revanna, A. (2022a). Arbuscular Mycorrhizal Fungi Influence Crop Productivity, Plant Diversity, and Ecosystem Services. In: Rajpal, V.R., Singh, I., Navi, S.S. (eds) Fungal diversity, ecology and control management. *Fungal Biol*, Springer, Singapore, pp 345-362. https://doi.org/10.1007/978-981-16-8877-5_16.

Bagyaraj, D. J., Leena, S., Nikhil Sai, N., and Ashwin, R. (2022b). Inoculation with selected microbial consortia promotes growth of chilli and basil seedlings raised in pro trays. *J Microbiol Biotechnol*, 7, 21-27.

Bagyaraj, D. J., Praveen, R., and Ashwin, R. (2022c). Arbuscular mycorrhizal fungi: Role in sustainable agriculture. *Biofertilizer Newsletter*, 30, 12-22.

Bagyaraj, D. J., and Mallesha, B. C. (2000). Potentials for the use of VA mycorrhiza in horticulture. In: Aneja KR, Charaya MD, Aggarwal A, Hans DK, Khan SA, editors. *Glimpse in Plant Sciences*. Meerut (India): Pragathi Prakashan. 8, 37-41.

Bagyaraj, D. J., Sharma, M. P., and Maiti, D. (2015). Phosphorus nutrition of crops through arbuscular mycorrhizal fungi. *Cur Sci*, 108, 1288-1293.

Bagyaraj, D. J., and Sturmer, S. L. (2008). Arbuscular Mycorrhizal Fungi (AMF). In: "*A Handbook of Tropical Soil Biology: Sampling and Characterization of Belowground Biodiversity*", (Eds.): Moreira, F. M. S., Huising, J. E. and Bignell, D. E. Earthscan Publication, London, 131-147.

Bagyaraj, D. J. (2011). *Microbial Biotechnology for Sustainable Agriculture, Horticulture and Forestry*. New Delhi: New India Publishing Agency.

Bagyaraj, F. J., Manjunath, A., and Govinda, Y. S. (1988). Mycorrhizal inoculation effect on different crops. *J Soil Biol Ecol*, 8, 98-103.

Bahadur, A., Batool, A., Nasir, F., Jiang, S., Mingsen, Q., Zhang, Q., Pan, J., Liu, Y., and Feng, H. (2019). Mechanistic insights into arbuscular mycorrhizal fungi-mediated drought stress tolerance in plants. *Int J Mol Sci*, 20(17), 41-59.

Begum, N., Qin, C., Ahanger, M. A., Raza, S., Khan, M. I., Ashraf, M., Ahmed, N., and Zhang, L. (2019). Role of arbuscular mycorrhizal fungi in plant growth regulation: implications in abiotic stress tolerance. *Front Plant Sci*, 10, 1068-1073.

Benedetto, A., Magurno, F., Bonfante, P., and Lanfranco, L. (2005). Expression profiles of a phosphate transporter gene (GmosPT) from the endomycorrhizal fungus *Glomus mosseae*. *Mycorrhiza*, 15, 620-627.

Berger, F., and Gutjahr, C. (2021). Factors affecting plant responsiveness to arbuscular mycorrhiza. *Curr Opin Plant Biol*, 59, 1019-1034.

Berruti, A., Lumini, E., Balestrini, R., and Bianciotto, V. (2016). Arbuscular mycorrhizal fungi as natural biofertilizers: let's benefit from past successes. *Front Microbiol*, 5(12), 13-22

Bona, E., Cantamessa, S., Massa, N., Manassero, P., Marsano, F., Copetta, A., Lingua, G., D'Agostino, G., Gamalero, E., and Berta, G. (2017). Arbuscular mycorrhizal fungi and plant growth-promoting pseudomonads improve yield, quality and nutritional value of tomato: a field study. *Mycorrhiza*, 27, 1-1.

Borde, M., Dudhane, M., and Jite, P. K. (2009). Role of bioinoculant (AM fungi) increasing in growth, flavor content and yield in *Allium sativum* L. under field condition. *Notulae Botanicae Horti Agrobotanici Cluj-Napoca*, 37(2), 124-138.

Bucher, M., Hause, B., Krajinski, F., and Küster, H. (2014). Through the doors of perception to function in arbuscular mycorrhizal symbioses. *New Phytol*, 204(4), 833-840.

Caruso, T., Mafrica, R., Bruno, M., Vescio, R., and Sorgonà, A. (2021). Root architectural traits of rooted cuttings of two fig cultivars: Treatments with arbuscular mycorrhizal fungi formulation. *Sci Hortic*, 283, 1100-1183.

Cely, M. V., De Oliveira, A. G., De Freitas, V. F., De Luca, M. B., Barazetti, A. R., Dos Santos, I. M. O., Gionco, B., Garcia, G. V., Prete, C. E. C., and Andrade, G. (2016). Inoculant of arbuscular mycorrhizal fungi (*Rhizophagus clarus*) increase yield of soybean and cotton under field conditions. *Front Microbiol*, 7, 720.

Chifetete, V. W., and Dames, J. F. (2020). Mycorrhizal interventions for sustainable potato production in Africa. *Front Sustain Food Syst*, 4, 1-17, doi: 10.3389/fsufs.2020.593053

Coelho, I. R., Pedone-Bonfim, M. V. L., Silva, F. S. B., and Maia, L. C. (2014). Optimization of the production of mycorrhizal inoculum on substrate with organic fertilizer. Brazilian *J Microbiol*, 45, 1173–1178.

Colizzi, L., Caivano, D., Ardito, C., Desolda, G., Castrignanò, A., Matera, M., Khosla, R., Moshou, D., Hou, K. M., Pinet, F., and Chanet, J. P. (2020). Introduction to agricultural IoT. In: *Agricultural Internet of Things and Decision Support for Precision Smart Farming*. Academic Press. 1-33.

Crossay, T., Antheaume, C., Redecker, D., Bon, L., Chedri, N., Richert, C., Guentas, L., Cavaloc, Y., and Amir, H. (2017). New method for the identification of arbuscular mycorrhizal fungi by proteomic-based biotyping of spores using MALDI-TOF-MS. *Scientific Reports*, 7, 1-16.

Dames, J. F., and Ridsdale, C. J. (2012). What we know about arbuscular mycorrhizal fungi and associated soil bacteria. *African J Biotechnol*, 11, 13753-13760.

Douds, D. D., Nagahashi, G., Reider, C., and Hepperly, P. R. (2007). Inoculation with arbuscular mycorrhizal fungi increases the yield of potatoes in a high P soil. *Biological Agri Horti*, 25(1), 67-78.

Dsouza, J. (2019). Techniques for the mass production of Arbuscular Mycorrhizal fungal species. In: *Advances in Biological Science Research*. Academic Press, 445-451.

Emmanuel, O. C., and Babalola, O. O. (2020). Productivity and quality of horticultural crops through co-inoculation of arbuscular mycorrhizal fungi and plant growth promoting bacteria. *Microbiol Res*, 239, 1265-1269.

Estaún, V., Camprubí, A., and Joner, E. J. (2002). Selecting arbuscular mycorrhizal fungi for field application. In: Gianinazzi S, Schüepp H, Barea JM, Haselwandter K, Eds. Mycorrhizal Technology in Agriculture: From Genes to Bioproducts. *Birkhäuser Verlag*, 249-259.

Faghihinia, M., Jansa, J., Halverson, L. J., and Staddon, P. L. (2023). Hyphosphere microbiome of arbuscular mycorrhizal fungi: a realm of unknowns. *Biol Fertility Soils*, 59(1), 17-34.

Fall, A. F., Nakabonge, G., Ssekandi, J., Founoune-Mboup, H., Apori, S. O., Ndiaye, A., Badji, A., and Ngom, K. (2022). Roles of arbuscular mycorrhizal fungi on soil fertility: Contribution in the improvement of physical, chemical, and biological properties of the soil. *Front Fungal Biol*, https://doi.org/10.3389/ffunb.2022.723892

Fortin, J. A., Bécard, G., Declerck, S., Dalpé, Y., St-Arnaud, M., Coughlan, A. P., and Piché, Y. (2002). Arbuscular mycorrhiza on root-organ cultures. *Can J Bot*, 80, 1-20.

Frąc, M., Hannula, S. E., Bełka, M., and Jędryczka, M. (2018). Fungal biodiversity and their role in soil health. *Frontiers Microbiol*, 13(9), 707.

Frank, A. B. (1885). Über die auf Wurzelsymbiose beruchende Ernarung gewisser Baume durch unterirdische Pilze. *Ber Deutch Bot Gessell* [On the nutrition of certain trees, based on root symbiosis, by subterranean fungi. *About Deutch Bot Gessell*], 3, 128-145.

Garg, N., Singh, S., and Kashyap, L. (2017). Arbuscular mycorrhizal fungi and heavy metal tolerance in plants: an insight into physiological and molecular mechanisms. In: *Mycorrhiza-nutrient uptake, biocontrol, ecorestoration*. Springer, 75-97.

Geetha, K., and Dathar, V. (2022). Plant–Fungal Interactions. In *Applied Mycology: Entrepreneurship with Fungi*, 15, 271-285). Cham: Springer International Publishing.

Gerdemann, J. W. (1975). Vesicular arbuscular mycorrhizae. In: Torrey JC, Clarkson DT, editors. *The Development and Function of Roots*. New York (NY): Academic Press, 575-591.

Giovannini, L., Sbrana, C., Giovannetti, M., Avio, L., Lanubile, A., Marocco, A., and Turrini, A. (2022). Diverse mycorrhizal maize inbred lines differentially modulate mycelial traits and the expression of plant and fungal phosphate transporters. *Sci Rep*, 12(1), 212-279.

Gupta, R. P., Kalia, A., and Kapoor, S. (2007). *Bioinoculants – A Step Towards Sustainable Agriculture*. New Delhi: New India Publishing Agency.

Haro, H., Semdé, K., Bahadio, K., and Ganaba, S. (2020). Mycorrhizal inoculation effect on the forage cowpea biomass production in Burkina Faso. *American J Plant Sci*, 11(11), 1714-1722.

Hung, L. L., and Sylvia, D. M. (1988). Production of vesicular-arbuscular mycorrhizal fungus inoculum in aeroponic culture. *Appl Environ Microbiol*, 54, 353-357.

Ijdo, M., Cranenbrouck, S., and Declerck, S. (2011). Methods for large-scale production of AM fungi: Past, present, and future. *Mycorrhiza*, 21, 1–16.

Jyothi, E., and Bagyaraj, D. J. (2018). Microbial consortia developed for *Ocimum tenuiflorum* reduces application of chemical fertilizers by 50% under field conditions. *Med. Plants – Int J Phytomed Related Ind*, 10, 138-144

Kokkoris, V., and Hart, M. (2019). *In vitro* propagation of arbuscular mycorrhizal fungi may drive fungal evolution. *Front Microbiol*, 10, 24-31.

Kormanik, P. P., Bryan, W. C., and Schultz, R. C. (1980). Procedure and equipment for staining large number of plant root samples for endomycorrhizal assay. *Can J Microbiol*, 26, 536-538.

Kuila, D., and Ghosh, S. (2022). Aspects, problems and utilization of Arbuscular Mycorrhizal (AM) application as bio-fertilizer in sustainable agriculture. *Curr Res Microb Sci*, 6, 100-107.

Lakshmipathy, R., Balakrishna, A. N., Bagyaraj, D. J., and Kumar, D. P. (2000). Symbiotic response of cashew root stocks to different VA mycorrhizal fungi. *Cashew*, 14(3), 20-4.

Li, Y. W., Tong, C. L., and Sun, M. F. (2022). Effects and Molecular Mechanism of Mycorrhiza on the Growth, Nutrient Absorption, Quality of Fresh Leaves, and Antioxidant System of Tea Seedlings Suffering from Salt Stress. *Agronomy*, 12(9), 2163-2174.

Lovato, P., Guillemin, J. P., and Gianinazzi, S. (1992). Application of commercial arbuscular endomycorrhizal fungi inoculants to the establishment of micropropagated grapevine rootstock and pineapple plants. *Agron J*, 12, 873-880.

Mahanty, T., Bhattacharjee, S., Goswami, M., Bhattacharyya, P., Das, B., Ghosh, A., and Tribedi, P. (2017). Biofertilizers: a potential approach for sustainable agriculture development. *Environ Sci Pollut Res*, 24, 3315-3335.

Malviya, D., Singh, P., Singh, U. B., Paul, S., Kumar Bisen, P., Rai, J. P., Verma, R. L., Fiyaz, R. A., Kumar, A., Kumari, P., Dei, S., Ahmed, M. R., Bagyaraj, D. J., and Singh, H. V. (2023). Arbuscular mycorrhizal fungi-mediated activation of plant defense responses in direct seeded rice (*Oryza sativa* L.) against root-knot nematode *Meloidogyne graminicola*. *Front Microbiol*, 14, 110-141. DOI: 10.3389/fmicb.2023.1104490.

IJdo, Marleen, Sylvie Cranenbrouck, and Stéphane Declerck. (2011). "Methods for large-scale production of AM fungi: past, present, and future." *Mycorrhiza*, 1-16.

Maússe Sitoe, S. D. N., and Dames, J. F. (2022). Mitigating Climate Change: The Influence of Arbuscular Mycorrhizal Fungi on Maize Production and Food Security. In *Arbuscular Mycorrhizal Fungi in Agriculture* (de Sousa RN, editor). Ch 4, Intech Open DOI:10.5772/intechopen.107128.

Menge, J. A., Johnson, E. L. V., and Minassian, V. (1979). Effect of heat treatment and three pesticides upon the growth and production of the mycorrhizal fungus *Glomus fasciculatum*. *New Phytol*, 82, 473-480.

Mercy, M. A., Shivashankar, G., and Bagyaraj, D. J. (1990). Mycorrhizal colonization in cowpea is host dependent and heritable. *Plant Soil*, 121, 292-294

Muleta, D. (2017). Legume response to arbuscular mycorrhizal fungi inoculation in sustainable agriculture. Springer International Publishing.

Muthukumar, T., Sumathi, C. S., Rajeshkannan, V., and Bagyaraj, D. J. (2022). Mycorrhizosphere Revisited: Multitrophic Interactions. In: Singh, U.B., Rai, J.P., Sharma, A.K. (eds) Re-visiting the Rhizosphere Eco-system for Agricultural Sustainability. *Rhizosphere Biol*, 17, 9-35. https://doi.org/10.1007/978-981-19-4101-6_2.

Mythra, R., Srikantha, G. S., Jagadeesh, U., Bhagyashree, K. B., and Manjunath, A. (2022). Effect of AM fungus on phosphorus nutrition of maize and pigeon pea in alfisols as influenced by different phosphorus amendments of North Carolina Rock Phosphate (NCRP). *Kavaka*, 58, 28-35.

Nunes, J. L. (2018). Anatomic and Morphologic Relations Developed Between Plants Roots and Arbuscular Mycorrhizal Fungi (AFM) During the Establishment of Mycorrhizal Symbiosis: A Review. *Open Acc J Agri Res*, 3(02), 55-65.

Omar, A. E. (2007). Rooting and growth response of grapevine nurslings to inoculation with arbuscular mycorrhizal fungi and irrigation intervals. *J Appl Hortic*, 9(2), 108-11.

Ortas, I. (2019). Comparison of indigenous and selected mycorrhiza in terms of growth increases and mycorrhizal dependency of sour orange under phosphorus and zinc deficient soils. *Eur J Hortic Sci*, 84, 218-225.

Plenchette, C., Furlan, V., and Fortin, J. A. (1983). Growth stimulation of apple trees in unsterilized soil under field conditions with mycorrhizal inoculation. *Can J Bot*, 61(8), 2003-2008.

Qin, Z., Peng, Y., Yang, G., Feng, G., Christie, P., Zhou, J., Zhang, J., Li, X., and Gai, J. (2022). Relationship between phosphorus uptake via indigenous arbuscular mycorrhizal fungi and crop response: A ^{32}P-labeling study. *Appl Soil Ecol*, 1, 1046-1054.

Raghu, H. B., Anuroopa, N., Ashwin, R., Harinikumar, K. M., Ravi, J. E., and Bagyaraj, D. J. (2021). Selected microbial consortia promotes *Dalbergia sissoo* growth in the large-scale nursery and wastelands in a semi-arid region in India. *J For Res*, https://doi.org/10.1080/13416979.2021.1955439.

Ranadev, P., Ashwin, R., Anuroopa, N., and Bagyaraj, D. J. (2022). Symbiotic response of fodder cowpea (*Vigna unguiculata* L.) and field bean (*Lablab purpureus* L.) with different arbuscular mycorrhizal fungi. *Kavaka*, 58(3), 34-38.

Ren, A. T., Zhu, Y., Chen, Y. L., Ren, H. X., Li, J. Y., Abbott, L. K., and Xiong, Y. C. (2019). Arbuscular mycorrhizal fungus alters root-sourced signal (abscisic acid) for better drought acclimation in Zea mays L. seedlings. *Environ Exp Bot*, 167, 103-124.

Rui, W., Mao, Z., and Li, Z. (2022). The Roles of Phosphorus and Nitrogen Nutrient Transporters in the Arbuscular Mycorrhizal Symbiosis. *Int J Mol Sci*, 23(19), 110-127.

Seemann, C., Heck, C., Voß, S., Schmoll, J., Enderle, E., Schwarz, D., and Requena, N. (2022). Root cortex development is fine-uned by the interplay of MIGs, SCL3 and DELLAs during arbuscular mycorrhizal symbiosis. *New Phytol*, 233(2), 948-965.

Selvakumar, G., Krishnamoorthy, R., Kim, K., and Sa, T. (2016). Propagation technique of arbuscular mycorrhizal fungi isolated from coastal reclamation land. *European J Soil Biol*, 74, 39-44.

Shashikala, B. N., Reddy, B. J. D., and Bagyaraj, D. J. (1999). Influence of *Glomus mosseae* on the growth of micropropagated Syngonium and Spathiphyllum at varied levels of fertilizer phosphorus. *Crop Res*, 18, 900-912.

Sheeba, J. J., Ranadev, P., Ashwin, R., and Bagyaraj, D. J. (2022). Influence of AM fungus *Funneliformis mosseae* and K solubilizing bacterium *Bacillus mucilaginosus* on the growth of tomato seedlings raised in pro trays. *J Microbes Res*, 1(2), 1-6.

Singh, M., Rakshit, R., and Beura, K. (2020). Endomycorrhizal Fungi: Phosphorous Nutrition in Crops. In: *Sustainable Agriculture*. Apple Academic Press, 203-209.

Smith, S., and Dickson, S. (1997). VA Mycorrhizas: Basic Research Techniques. Adelaide, Australia.

Spoorthi, V. B., Ranadev, P., Ashwin, R., and Bagyaraj, J. D. (2022). Response of *Capsicum annuum* L. Seedlings Raised in Pro Trays to Inoculation with AM Fungus *Glomus bagyarajii* and K Solubilizing Bacterium *Frateuria aurantia*. *Seeds*, 1, 315-323.

Sun, R. T., Zhang, Z. Z., Liu, M. Y., Feng, X. C., Zhou, N., Feng, H. D., Hashem, A., Abd_Allah, E. F., Harsonowati, W., and Wu, Q. S. (2022). Arbuscular mycorrhizal fungi and phosphorus supply accelerate main medicinal component production of *Polygonum cuspidatum*. *Frontiers Microbiol*, 13, 354-355.

Tarbell, T. J., and Koske, R. E. (2007). Evaluation of commercial arbuscular mycorrhizal inocula in a sand/peat medium. *Mycorrhiza*, 18, 51-56.

Tawaraya, K. (2003). Arbuscular mycorrhizal dependency of different plant species and cultivars. *Soil Sci Plant Nut*, 49(5), 655-668.

Tenzin, U. W., Noirungsee, N., Runsaeng, P., Noppradit, P., and Klinnawee, L. (2022). Dry-Season Soil and Co-Cultivated Host Plants Enhanced Propagation of Arbuscular Mycorrhizal Fungal Spores from Sand Dune Vegetation in Trap Culture. *J Fungi*, 8(10), 1061-1073.

Thilagar, G., Bagyaraj, D. J., Podile, A. R., and Vaikuntapu, P. R. (2016). Bacillus sonorensis, a novel plant growth promoting rhizobacterium in improving growth, nutrition and yield of chilly (*Capsicum annuum* L.). *Proc Natl Acad Sci India Sect B Biol Sci*, 88, 813-818.

Thilagar, G., and Bagyaraj, D. J. (2015). Influence of different arbuscular mycorrhizal fungi on growth and yield of chilly. *Proc Natl Acad Sci India Sect B Biol Sci*, 85, 71-75.

Thilagar, G., Bagyaraj, D. J., and Rao, M. S. (2015). Selected microbial consortia developed for chilly reduces application of chemical fertilizers by 50% under field conditions. *Scientia Hort*, 198, 27-35.

Tilak, K. V. B. R., Pal, K. K., and Dey, R. (2010). *Microbes for Sustainable Agriculture*. New Delhi: I. K. International Publishing House Pvt. Ltd.

Trejo-Aguilar, D., and Banuelos, J. (2020). Isolation and culture of arbuscular mycorrhizal fungi from field samples. Arbuscular Mycorrhizal Fungi: *Methods and Protocols*, 1-8.

Wang, D., Dong, W., Murray, J., and Wang, E. (2022). Innovation and appropriation in mycorrhizal and rhizobial symbioses. *Plant Cell*, 34(5), 1573-1599.

Weng, W., Yan, J., Zhou, M., Yao, X., Gao, A., Ma, C., Cheng, J., and Ruan, J. (2022). Roles of arbuscular mycorrhizal fungi as a biocontrol agent in the control of plant diseases. *Microorganisms*, 10, 1266-1276.

Wright, S. F. (2005). Management of arbuscular mycorrhizal fungi. *Roots and soil management: interactions between roots and the soil*, 48, 181-97.

Yang, Y., He, C., Huang, L., Ban, Y., and Tang, M. (2017). The effects of arbuscular mycorrhizal fungi on glomalin-related soil protein distribution, aggregate stability and their relationships with soil properties at different soil depths in lead-zinc contaminated area. *PloS one.*, 12(8), 182-194.

Zhang, X., Gao, H., Liang, Y., and Cao, Y. (2021). Full-length transcriptome analysis of asparagus roots reveals the molecular mechanism of salt tolerance induced by arbuscular mycorrhizal fungi. *Environ Exp Bot*, 185, 104-122.

Zhu, Y. G., and Smith, S. E. (2001). Seed phosphorus (P) content affects growth, and P uptake of wheat plants and their association with arbuscular mycorrhizal (AM) fungi. *Plant and Soil*, 231, 105-112.

Biographical Sketch

Name: Dr. Davis Joseph Bagyaraj

Affiliation: Centre for Natural Biological Resources and Community Development (CNBRCD)

Education: Ph.D. (Agriculture Microbiology)

Business Address: CNBRCD, 41 RBI Colony, Anand Nagar, Bengaluru -5260024

Research and Professional Experience: Prof. D. Joseph Bagyaraj earned his B.Sc. (Agri.) from Agricultural College, Bangalore and Ph.D. in Agri. Microbiology from UAS, Bangalore. He joined the UAS, Bangalore as an Assistant Professor of Microbiology in 1966 and then became Professor and Head. He had his post-doctoral training at New Zealand, Australia, UK and USA. He also worked as a Visiting Scientist at New Zealand and as a Senior Scientist at Oregon State University, USA. Currently he is NASI Hon. Scientist and Chairman, Center for Natural Biological Resources and Community Development, Bangalore.

Initiated systematic investigations on arbuscular mycorrhizal fungi (AMF) in India and is considered to be a pioneering scientist in this area of research. He has developed facile procedures for screening and selecting AMF for crops important in agriculture, horticulture and forestry. His studies demonstrated 25-50% saving of P fertilizer through AMF inoculation. He has published 425 research papers, 114 review articles and written 11 books. He

Arbuscular Mycorrhizal Fungi as Biofertilizers ... 31

has taught Microbiology to UG and PG students for the past 50 years and has mentored nearly 65 M.Sc. and Ph.D. students.

Dr. Bagyaraj has been invited either to deliver lectures or evaluate projects on mycorrhiza in different parts of the world. He has trained scientists from several countries on mycorrhizal techniques. He is a Fellow of the most of the National Academies in the country. He is currently the President of the Indian Society of Soil Biology and Ecology. He is/was also an Expert Committee Member of several funding agencies like ICAR, DST, DBT, CSIR, UGC, NBA, IFS (Sweden), AICAR (Australia), UNEP and FAO.

Professional Appointments: Formerly Professor and Head, Department of Microbiology, University of Agricultural Sciences, Bangalore – 560065

Honors: Dr. Bagyaraj received several awards and honours; to name a few, Shome Memorial Award; SR Vyas Memorial Award; Rangaswami Memorial Award; Prof. T.S. Sadasivan Lecture Award; Dr. V Agnihotrudu Memorial Lecture Award; Distinguished Asian Mycologist Award, 2019 conferred in Japan and many others. He is also currently involved as Convenor/Resource person of the National Academy of Sciences taking recent advances in Agricultural Microbiology to college students and teachers. He was invited by the European Commission to contribute on mycorrhizal fungi for the Global Atlas on Soil Biodiversity (only Indian scientist invited) which was released at Belgium in 2016. Honouring his contributions a mycorrhizal fungus has been named as ***Glomus bagyarajii.***

Publications from the Last 3 Years: Research Papers – 18; Book Chapters - 08

Chapter 2

Zinc Solubilizing Bacteria: An Emerging Biofertilizer for Sustainable Agriculture

Mohammad Reza Sarikhani[1,*], PhD
Bahman Khoshru[2], PhD
and Md Shafiul Islam Rion[3], PhD

[1]Department of Soil Science, Faculty of Agriculture, University of Tabriz, Iran
[2]Soil and Water Research Institute, Agricultural Research,
Education and Extension Organization (AREEO), Karaj, Iran
[3]Division of Plant and Soil Sciences, West Virginia University, West Virginia, US

Abstract

Recently special attention has been paid to the use of biological potentials as environmentally friendly agents and replacing them with chemical compounds. To achieve sustainable agriculture and use eco-friendly approaches, soil microbial potentials and its metabolites can be used to supply necessary plant elements such as zinc (Zn). Zinc is a key micronutrient required for growth by all living forms, including plants, humans, and microorganisms. Humans and other living organisms need a small amount of zinc in their lives for proper physiological functions. Zinc is a vital micronutrient for plants that performs various important functions in their life cycle. Zinc deficiency is a known problem that occurs in both plants and humans. The use of zinc-solubilizing rhizobacteria can be a sustainable intervention to increase the bioavailability of zinc in the soil, which can be useful in reducing yield loss and zinc malnutrition. Newly, several reports have been published on the use of soil bacteria in converting insoluble Zn into plant-available

---

[*] Corresponding Author's Email: rsarikhani@yahoo.com.

In: Biofertilizers
Editor: Philip L. Bevis
ISBN: 979-8-89113-082-1
© 2023 Nova Science Publishers, Inc.

and soluble forms to eliminate Zn deficiency. Bacterial genera such as *Pseudomonas*, *Bacillus*, *Acinetobacter*, *Azotobacter*, *Azospirillum*, *Gluconacetobacter*, *Burkholderia*, and *Thiobacillus* have shown their ability to solubilize Zn through different mechanisms such as secreting organic acids, siderophores, and other chelating compounds. In addition, these Zn solubilizing bacteria (ZSB) have also demonstrated the ability to improve crop quality via PGP traits including phytohormone production, phosphate solubilization etc. Although some studies have been recently done in isolation and identification of ZSB in sustainable agriculture, mechanisms of action and prospects of these rhizobacteria in sustainable agriculture stay to be completely caught on. This study principally focuses on the ZSB and assesses their diversity, action mechanisms, ecological significances and its biotechnological applications. Such data is valuable in deciding their potential and assessing their prospects in promoting sustainable agricultural systems. The utilize of ZSB may be a promising methodology to create and fulfill Zn demand of the growing crop without causing any environmental hazard. This review critically examines the current state-of-art on use of ZSB strains as biofertilizers and the important roles performed by these beneficial microbes in maintaining soil fertility and enhancing crop productivity.

Keywords: biofertilizer, PGPR, sustainable agriculture, zinc (Zn), Zn solubilization

Introduction

Zinc (Zn) is a key micronutrient, required for all living forms including plants, humans, and microorganisms for their development. Humans and other living organisms require zinc in their lives in little amounts for proper physiological functions. Zinc is a crucial micronutrient for plants which plays various important functions in their life cycle (Bonaventuraa et al., 2015).

Zn plays a role in the metabolism of carbohydrates and auxin in plants (Alloway, 2008). It is also considered an antioxidant agent in plants (Alloway, 2004). Symptoms of Zn deficiency in plants include chlorosis, reduced stem growth, reduced leaf size, sensitivity to diseases and fungal infections, sensitivity to heat, and it also has a negative effect on root growth, water absorption and transfer, pollen formation, and seed yield (Tavallali et al., 2010). Zn can be absorbed by plants in the form of divalent cation, but a very small part of the total Zn enters the soil solution, and most of it exists in the form of sediment and insoluble complexes in the soil, which cannot be absorbed by plants and leads to the manifestation of Zn deficiency symptoms as one of the most common micronutrient deficiencies in plants (Kabata-Pendias and Pendias, 2001; Alloway, 2008).

To deal with Zn deficiency in plants, farmers use chemical fertilizers such as Zn sulfate (White and Broadly, 2005) or Zn-EDTA (Karak et al., 2005). With these fertilizers, Zn introduced into the soil turns into insoluble complex forms within 7 days, which will lead to environmental damage (Rattan and Shukla, 1991). To increase Zn absorption by plants, in different areas, strategies such as crop rotation and regular intercropping (Zuo and Zhang, 2009), transgenic and genetic engineering approaches (Tan et al., 2015) and conventional breeding (Cakmak et al., 2010) have been used. However, these approaches are expensive, laborious and slower. A better alternative to all these approaches is the use of plant growth promoting bacteria (PGPB) (Khoshru et al., 2022).

PGPB or plant growth promoting rhizobacteria (PGPR) are a group of soil-borne bacteria that multiply and grow in the soil, and by colonizing the roots of various plants, they stimulate their growth and development (Khoshru et al., 2020c). Helping to increase plant growth by these bacteria is done directly or indirectly (Sarikhani et al., 2019a; Khoshru et al., 2023). PGPR increase the growth and performance of their host plants with mechanisms such as increasing the bioavailability of elements via activities such as nitrogen fixation, phosphorus solubilization, potassium release from minerals, solubilization of other elemets such as Fe, Zn, and Mn, siderophore production (increasing the availability of micronutrients), phytohormone production, production of various organic acids, proton secretion, production of ACC deaminase enzyme, production of volatile organic compounds, various antibiotics and control of plant pathogens, etc. Zn solubilizing bacteria (ZSB) are a group of PGPR that increase the bioavailability of Zn by mechanisms such as pH reduction due to the production of various organic acids, proton secretion, siderophore production, etc. (Khoshru et al., 2022 and 2023; Sarikhani et al., 2020). ZSB has the ability to solubilize Zn in compounds, complexes and insoluble sediments, and by solubilizing these compounds and increasing the availability of Zn for plant roots, they facilitate the absorption of this element by plants and thereby increase their growth (Khoshgoftarmanesh et al., 2018).

Considering the problems of long-term use of chemical Zn fertilizers and the damage caused to the soil and the environment, along with the economic problems for the farmer, the need to use alternative strategies to supply this element is felt more. The approach of using the potential of ZSB to supply this element is an environmentally friendly solution in line with sustainable agriculture. Accordingly, in this review, it has been tried to discuss on the status of Zn in soil and plants and various factors affecting its mobility or deposition, and further focus on the potential of ZSM (Zn solubilizing microorganisms) especially ZSB to solubilize this element in the soil and increase its absorption by plants. We have a look to their mechanism of action, their diversity and future prospect in this area.

Zn in Soils and Its Availability

The chemical composition of the parent rock and its weathering rate determine the availability of Zn in unpolluted and unfertilized soils (Chesworth, 1991). The amount of Zn in sedimentary rocks is from 80 to 120 mg/kg, in igneous rocks from 40 to 120 mg/kg, in limestone, sandstone and dolomites from 10 to 25 mg/kg and in argillaceous sediments and shales up to 15-30 mg/kg is variable (Kabata-Pendias, 2000). Zn in agricultural soils varies from 10 to 300 mg/kg, and the total Zn content in these soils tends to average 50-65 mg/kg according to various studies (Barber, 1995). Zn has the lowest amount in sandy soils and the highest amount in calcareous soils (Figure 1) (Kabata-Pendias, 2000). Some researchers have reported an average of 64 mg/kg Zn for soils worldwide (Kabata-Pendias & Pendias, 1999).

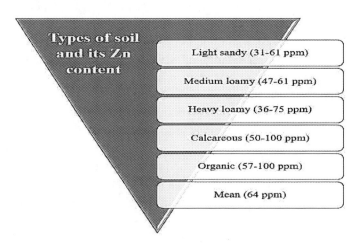

Figure 1. Abundance of Zn in different soils (Kabata-Pendias, 2000; Wolf et al., 2022).

As a factor that can reduce the yield by more than 40%, Zn deficiency is considered one of the most common geographical limitations of elements among micronutrients (Singh et al., 2005), most losses of crop yield related to Zn may be of human or geological origin (Singh et al., 2005; Sadeghzadeh, 2013). In calcareous soils with high pH, the loss of topsoil (rich in micronutrients and organic matter) as a result of land leveling by machines for uniform irrigation will lead to micronutrient deficiency, especially Zn, in agricultural products (Alloway, 2008). Also, sandy soils (with low total Zn content) and soils with high phosphorus content have also shown Zn deficiency in crops (Singh et al., 2005). To find out the condition of the soil in terms of the capacity of providing Zn for the plant, the total Zn is not a reliable indicator because only a small part of the total Zn (<1 mg/kg) is present in the soil solution, which can be absorbed by the plant (Kabata-Pendias, 2000).

In arid and semi-arid regions, Zn deficiency is observed in the upper layers of the soil (which are drier) to a greater extent compared to the deeper layers of the soil (which are more humid). In dry soil layers, the diffusion of soil Zn to plant roots decreases (Cakmak et al., 1996). In such a case, the use of liquid Zn fertilizers and its injection into the lower layers has a better result compared to the surface use of fertilizer, although the cost of liquid fertilizers is higher, but it prevents the occurrence of Zn deficiency symptoms for a significant period of time (Holloway, 1996).

Soil-related factors are the main cause of Zn deficiency in agricultural products, factors such as soil calcareousness, high pH (pH>7), sodic soil, light soil texture, lack of organic matter, antagonistic relationship of elements (such as high iron/manganese oxide contents, high calcium carbonate and bicarbonate contents, high sodium and phosphorus availability, high exchangeable magnesium/calcium ratio), soil compaction and limitation of root expansion, high water level and waterlogging, etc. (Cakmak, 2008; Alloway, 2009). According to the above, it can be seen that Zn deficiency can occur in a wide range of soils. Related to the inheritance of soil, parent materials can also be another factor of Zn deficiency in soils. It has been reported that 92.8% of the acid soils studied in Northern Greece are deficient in Zn (Noulas et al., 2018). The amount of Zn recorded in the plants of the Greek island of Crete was between 0.2 and 12.3 mg/kg, which was attributed to the difference in parent materials and agricultural practices (Vavoulidou et al., 2009).

Zn is found in soil in 5 forms (pools), which include water-soluble Zn, exchangeable Zn, adsorbed Zn, chelated Zn and complexed Zn (Mandal and Mandal, 1986). The sensitivity to absorption and leaching of Zn in these forms is different and the balance between different pools of Zn can be affected by the pH and concentration of Zn and other metals, especially iron and manganese (Sadeghzadeh and Rengel, 2011). The content of Zn in the soil solution is controlled by absorption mechanisms, which plays a fundamental role in soil-plant relations, so that there is always some amount of absorbable Zn in the soil solution that is available to agricultural products, although its amount can be variable (from low to much). Equation 1 shows the adsorption of exchangeable Zn cations in soil.

$$Zn^{2+} + M\text{-Soil} \leftrightarrows Zn\text{-Soil} + M^{2+} \qquad (1)$$

The concentration of Zn in the soil solution and its availability to plants is controlled by Zn adsorption and desorption reactions between the soil solution and solid phase, which is influenced by pH, organic matter and mineral elements of the soil (Lindsay, 1991; Catlett et al., 2002). Differentiation of Zn in the soil is often done according to its chemical binding characteristics with different fractions, such as exchangeable Zn, Zn bounded to organic matter, Zn bounded to carbonate, Zn bounded to Fe-Mn oxides and residual Zn (Khoshgoftarmanesh et al., 2018). Among the different forms of Zn, the most sensitive binding form involves exchangeable Zn, which has the closest correlation with Zn uptake by plants (Chahal et al., 2005; Li et al., 2007). Zn bounded to organic matter is also available to plants due to the increased cation exchange capacity by organic matter and the presence of exchange sites for Zn on the surfaces of soil particles (Khoshgoftarmanesh et al., 2018). Unlike exchangeable Zn and organic matter-bounded Zn, binding of Zn to carbonates or Fe-Me oxides reduces its bioavailability for plants (Shuman and Wang, 1997).

Zn in Plants and Its Deficiency

The amount of Zn in dry matter of most plants is 30 to 100 mg/kg, and amounts above 300 mg/kg are generally toxic for plants. Plants such as corn, rice, sorghum, beans, grapes and citrus fruits are sensitive to Zn deficiency (Stein et al., 2007). Zn in plants as a micronutrient plays an important role in various enzymes (in the catalytic part) (Fageria, 2002 & 2004). These enzymes play a role in protein synthesis, carbohydrate metabolism, and regulation of auxin synthesis, pollen formation and maintaining the integrity of cell membranes

(Fageria, 2004). There are reports that show that Zn is closely related to nitrogen metabolism and therefore protein synthesis decreases in Zn-deficient plants (Fageria, 2002 & 2004). By providing Zn in Zn-deficient plants, the state of plant protein synthesis improved (Zeng et al., 2021). It can be said that because Zn is a structural component of ribosomes, its deficiency leads to the breaking of ribosomes, which further affects protein metabolism (Castillo-González et al., 2018). It has also been reported that Zn deficiency leads to inhibition of RNA and protein synthesis because this element participates in the structure of enzymes involved in protein synthesis such as glutamate dehydrogenase, ATPase and ribonuclease (Hafeez et al., 2013). Zn-containing enzymes include aldehyde dehydrogenases, carbonate dehydratases, Zn-finger DNA-binding proteins, and Zn/copper (Cu) superoxide dismutase (SOD) (Sofo et al., 2018; Al Jabri et al., 2022). By neutralizing superoxide radicals, Zn-Cu-SOD protects membrane proteins and lipids against oxidation (Alatawi et al., 2022). Zn also increases cytochrome production and is required for seed growth (Karthika et al., 2018). The role of Zn in some physiological functions of plants such as hormonal regulation (production of the amino acid tryptophan as a precursor of the auxin hormone) and signal transmission by mitogen-activated protein kinases has been reported (Bhantana et al., 2021; Kaur and Garg, 2021; Ahmad et al., 2022). Zn ions can also reduce a variety of harmful effects on plant cells by deactivating some proteins by binding to their functional groups or displacing other cations from binding sites (Natasha et al., 2022). Therefore, proper absorption, transfer and distribution of Zn in plant cells and tissues is important for proper plant function (Zlobin, 2021).

The root system is severely affected by zinc deficiency, impairing water and nutrient uptake (Fageria, 2004). Due to the poor availability of Zn in agricultural soils worldwide, which leads to a decrease in crop production and nutritional quality, the lack of this element is becoming a serious agricultural concern (Zeng et al., 2021).

Symptoms such as yellowing or chlorosis of immature leaves, small leaves, and wilting of stems are related to Zn deficiency. When Zn deficiency is severe, growth restriction, wilting, rolling, and browning of older leaves will be observed (Mattiello et al., 2015; Zhao and Wu, 2017; Xie et al., 2019).

Against the stress of Zn deficiency, plants adopt special mechanisms, including changes in the architecture of the root system, establishing symbiosis with beneficial microbes such as arbuscular mycorrhizae or Zn solubilizing bacteria, strengthening the defense mechanism against oxidative stress caused by Zn deficiency, producing organic acids and phytosiderophore, etc. (Zhao and Wu, 2017). The above changes can improve the absorption and transport of Zn in the plant (Gong et al., 2020).

On the contrary, exposure of plants to high amounts of Zn leads to a variety of disturbances in biological and chemical systems (Bankaji et al., 2019; Sidhu et al., 2020). Due to the comparable ionic radius, Zn toxicity can lead to the deficiency of other nutrients by interfering with their absorption and transport in the plant (Bankaji et al., 2019). The damage of this disorder can affect the performance of photosynthesis, stomatal conductance, and many metabolic activities, and therefore it will reduce plant growth and its structural and functional stability (Chakraborty and Mishra, 2020).

Most plants cannot overcome the problem of Zn deficiency in their tissues by relying only on their own abilities, and it is necessary for plants to have a relationship with some microbes in their rhizosphere, and therefore, microbial intervention in Zn absorption is a fact that has been pointed out in various researches.

Microbes-Mediated in Zn Solubilization

It is believed that the importance of microbial mediation in the soil is very vital for the permanent nutrition of agricultural systems. But it must be considered that there is still no complete understanding of environmental processes and related activities of soil microbes (Khoshru et al., 2020b). The most microbial activity and mediation in plant nutrition occurs in the rhizosphere region, and it is the region that is affected by plant root secretions and various microbial activities (Berendsen et al., 2021). Some rhizosphere microbes have a positive effect on plant growth and development by colonizing the root and through direct and indirect mechanisms (Kumar et al., 2017). In the imbalance nutritional condition, these microbes can increase nutrient absorption by increasing their bioavailability (for example, by solubilizing P, Zn, Fe, etc.). There are evidences that soil sterilization has led to the occurrence of signs of deficiency of elements such as Fe and Zn in plants, which have been attributed to soil microbe's activities (Jin et al., 2014; Ochieno, 2022). Shaikh and Saraf (2017) have reported that the inoculation of *Triticum aestivum* with PGPR strain *Exiguobacterium auranticum* MS-ZT10 increased the levels of Fe and Zn in the seed by 6 times. Fahsi et al. (2021) also reported that the presence of *Bacillus halotolerant* (J143), *Pseudomonas frederiksbergensis* (J158), and *Enterobacter hormaechei* (J146) strains in wheat rhizosphere led to a significant increase in Zn absorption and wheat germination.

The increase of Zn content in the roots and shoots of plants with the involvement of microbes especially bacteria has been reported many times, in which the inoculation of plants with PGPR has led to increased plant growth and yield, and improved its nutritional condition (Tariq et al., 2007; Khoshru et al., 2020b; Khoshru et al., 2023). These bacteria, which mainly belong to the genera *Pseudomonas*, *Rhizobium*, *Bacillus*, *Acrobactrum*, *Azospirillum*, *Azotobacter*, *Enterobacteriaceae*, *Stenotrophomonas*, and *Serratia*, are sometimes formulated as Zn biofertilizers in this field (Maleki et al., 2011). In addition to providing some nutrients required by the plant (one or more elements), these bacteria have led to the improvement of the overall performance of the plant (Oteino et al., 2015). Ramesh et al. (2014) reported that the inoculation of wheat and soybean plants with *B. aryabhattai* led to an improvement in the absorption of elements, especially Zn. Also, Lefèvre et al. (2014) reported a 7-12% increase in Zn in wheat grains by inoculation with certain strains of *Serratia* sp., *Bacillus* sp. *Pseudomonas* sp. In this study, in addition to increasing biomass, ZSB significantly increased shoot and root Zn content compared to non-inoculated plants. The implementation of successful plant-microbe interactions such as mineralization, solubilization and induction of physiological processes can be the reason for this ability of rhizobacteria or endophytes to promote plant growth (Lucas et al., 2014; Wang et al., 2014). Research has shown that the presence of PGPR in the rhizosphere region has led to many morphological and physiological changes in the host plant (Chattha et al., 2017). Khan et al. (2015) reported that the effective colonization of plant roots by different genera of PGPR can significantly increase the solubilization efficiency of nutrients such as P and Zn. Furthermore, positive effect of PGPB inoculation to the plants in supplying other elements such as K, Fe was reported in other studies (Khoshru et al., 20.., Sarikhani et al., 2018; Ebrahimi et al., 2023).

Isolation and Screening of ZSB

In order to obtain rhizospheric ZSB, it is necessary to isolate them on general-growth media such as tris minimal salt medium (TMS) via the serial dilution method (Figure 2). 100 microliters of the final dilutions (dilutions 10^{-6}, 10^{-7}, 10^{-8} and 10^{-9}) are used in 3 replicates on solid culture medium. After the appearance of colonies, theovernight culture of each isolate (grown in NB culture medium) is loaded on TMS culture medium via dot culture and then incubated. TMS culture medium should contain insoluble sources of Zn, such as Zn-phosphate, Zn-carbonate, Zn-oxide, etc. These resources should be used separately and with a concentration of 0.1% in the culture medium (Dinesh et al., 2015). The initial screening is based on the growth pattern, phenotype and morphology of the bacterial colony and the formation of a clear halo around the colony (Khoshru et al., 2020a). The TMS-Agar solid medium containing 0.1% of poorly soluble Zn sources (Zn-oxide, Zn-carbonate and Zn-phosphate) can be used separately to assessment the semi-quantitative Zn solubility by the isolates (Figure 1). After culturing the isolates in the medium (dot culture), the cultures are incubated at 28°C for 12 days. Then the ratio of halo diameter to colony diameter (HD/CD) is recorded for different days, for example, 3, 5, 8 and 12 days after initial cultivation (Saravanan et al., 2007a).

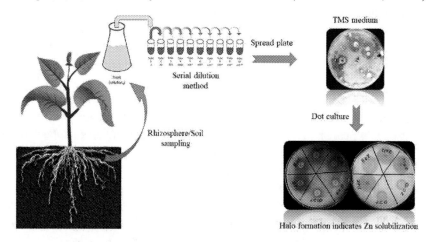

Figure 2. Schematic diagram of the ZSB isolation (Khoshru et al., 2023).

Although primary screening of ZSB is done in TSM solid medium containing different sources of insoluble zinc, final selection is based on their Zn solubilizing potential in liquid assay. In other words, efficiency of the selected bacteria from the first step should be assessed in TSM liquid media and the robust strains can be selected. For quantitative assessment of Zn solubilization, TMS-broth containing 0.1% of poorly soluble Zn sources is used. As mentioned earlier, insoluble sources of Zn should be used separately in TMS culture media. Next, 100μL of overnight suspension were inoculated into 50 ml of sterile TMS-broth media and kept in a shaker incubator at 28°C for 7 days with a rotation speed of 120 rpm. A TMS-broth medium without bacterial inoculation can be used for control treatment. After the incubation

Zinc Solubilizing Bacteria 43

time is over, the samples are removed from the incubator and centrifuged at 6000 rpm for 10 minutes, and finally the Zn concentration in the supernatants are measured using atomic absorption spectroscopy (Saravanan et al. 2007b). However, considering microbial biomass Zn can be considered too, as proposed by other researchers for estimation of K releasing efficiency of bacteria (Ebrahimi et al., 2019). To choose efficient ZSB strains, it is necessary to carry out a pot culture or field trial. We expect that there will be a positive correlation between their potentials either in laboratory used media or pot cultures. Attention to the other parameters such as pH reduction or siderophore production as a mechanism action of these bacteria can be considered in the study, too. More details are available in Khoshru et al. (2023).

Diversity of Zn-Solubilizers

A group of beneficial microorganisms that can solubilize inorganic Zn compounds from insoluble compounds is known as Zn-solubilizing microbes (ZSM). These microbes may have some other plant growth promoting traits beside Zn-solubilization. The list of ZSM is getting longer as more studies are happening in search of insoluble Zn-solubilizers in different ecosystems to support plant growth. The following table listed some of the ZSM found across the broad groups of bacteria, fungi, arbuscular mycorrhiza, actinomycetes, and cyanobacteria so far.

Table 1. Biodiversity of microorganisms having Zn solubilization potential

Microorganism	Instances	Source
Bacteria	*Pseudomonas fragi* *Pantoea dispersa* *Pantoea agglomerans* *Enterobacter cloacae*	Hashemnejad et al. (2021); Kamran et al. (2017)
	Acinetobacter sp.	Gandhi and Muralidharan (2016)
	Bacillus aryabhattai IA20 *Bacillus subtilis* IA6 *Paenibacillus polymyxa* IA7	Ahmad et al. (2021)
	Pseudomonas japonica	Eshaghi et al. (2019)
	Agrobacterium tumefaciens	Khanghahi et al. (2018)
	Ralstonia picketti *Pseudomonas aeruginosa* *Burkholderia cepacia* *Klebsiella pneumoniae*	Gontia-Mishra et al. (2017)

Table 1. (Continued)

Microorganism	Instances	Source
	Bacillus cereus	Batool et al. (2021); Khande et al. (2017); Shakeel et al. (2015)
	Bacillus megaterium	Bhatt and Maheshwari (2019)
	Bacillus altitudinis BT3 *Bacillus altitudinis* CT8	Kushwaha et al. (2021)
	Acinetobacter calcoaceticus *Bacillus proteolyticus* *Stenotrophomonas pavanii*	Sultan et al. (2023)
	Gluconacetobacter diazotrophicus	Intorne et al. (2009); Saravanan et al. (2007)
	Myristica yunnanensis *Stenotrophomonas chelatiphaga*	Ghavami et al. (2016)
	Burkholderia cenocepacia *Pseudomonas striata*	Pawar et al. (2015)
	Pseudomonas putida *Pseudomonas fluorescens*	Hashemnejad et al. (2021)
	Bacillus marisflavi	Kayalvizhi and Kathiresan (2019)
	Acinetobacter calcoaceticus *Agromyces italicus*	Khoshru et al. (2023)
	Paenibacillus polymyxa *Ochrobactrum intermedium* *Stenotrophomonas maltophilia* *Arthrobacter globiformi*	Batool et al., (2021)
Fungi	*Oidiodendron maius*	Martino et al. (2003)
	Aspergillus niger *Penicillium simplicissimum*	Nath et al. (2015); Sayer et al. (1995)
	Aspergillus fumigatus *Penicillium chrysogenum* *Penicillium crustosum* *Penicillium sclerotiorum*	Nath et al. (2015)
	Penicillium bilaji	Kucey (1988)
	Trichoderma longibrachiatum S12 *Trichoderma asperellum* S11 *Trichoderma atroviride* PHYTAT7	Sallam et al. (2021)
	Aspergillus chiangmaiensis *Aspergillus pseudopiperis* *Aspergillus pseudotubingensis*	Khuna et al. (2021)
	Trichoderma harzianum Rifai 1295-22	Altomare et al. (1999)
	Suillus luteus	Zhang et al. (2021)
	Fusarium haematococcum	Ravi et al. (2022)
	Astraeus odoratus *Phlebopus portentosus* *Pisolithus albus* *Scleroderma sinnamariense*	Kumla et al. (2014)
Microorganism	Instances	Source

Arbuscular mycorrhiza (AM)	*Glomus intraradices*	Giasson et al. (2005)
	Glomus mossae	Beura et al. (2016)
	Funneliformis mosseae	Chen et al. (2017)
	Rhizophagus intraradices	Balakrishnan and Subramanian (2016); Gupta et al. (2022)
	Rhizophagus irregularis	Watts-Williams and Cavagnaro (2018)

Bacteria as ZSB

Among Zn solubilizing microbes, rhizospheric bacteria have an important and special place (Shakeel et al., 2015; Khoshru et al., 2020a). ZSB plays a vital role in environmental cycle processes and solubilization of important and required plant elements such as Zn and iron, which significantly affects plant growth and development (Baig et al., 2012). Some rhizobacteria have a high ability to increase the bioavailability of Zn for plants and therefore have the potential to be used as Zn biofertilizers (Verma et al., 2015).

There are reports that show that the different strains of *Pseudomonas* as ZSB, in addition to the ability to solubilize Zn, can enhance plant growth through other mechanisms such as the production of phytohormones, phosphorus solubilization, siderophore production, antibiotic production, etc. (Reetha et al., 2014, Oteino et al., 2015). Pawar et al. (2015) reported that *Pseudomonas* sp. as a ZSB, it has anti-pathogenic properties by the mechanism of HCN production. Of course, there are conflicting reports in this connection, for example, it has been reported that the inoculation of wheat plant with *P. fragi* has led to an increase in nutrient absorption, germination rate and plant yield. These researchers attributed this growth enhancement to auxin production, phosphorus solubilization and HCN production, however, no report of Zn solubilization by *P. fragi* was found (Selvakumar et al., 2009). Another effective group of ZSB belongs to the *Pantoea* family, which have shown great abilities to enhance plant growth (Sarikhani et al., 2019b; Khoshru et al., 2020a). Majumdar and Chakraborty (2015) reported that *Pantoea agglomerans* by abilities such as effective solubilization of P and Zn, production of IAA, siderophore and also ammonia, led to a significant increase in weight, height, carbohydrate and chlorophyll content in barley plants. The same abilities have been reported for *Enterobacter asburiae* (Ahmad et al., 2022). It has been reported that *Rhizobium* sp. and *E. cloacae*, in addition to the ability to solubilize Zn, also produce plant hormones such as auxin and other bioactive compounds, and the inoculation of pea plants (*Pisum sativum*) by these strains has led to a significant increase in plant growth (Khalifa et al., 2016). Khoshru et al. (2023) in research obtained 20 isolates from the rhizosphere of corn, wheat and sunflower plants in different cities of East Azerbaijan province at Iran. Among the isolates, two isolates ZO11 and ZO14 were selected as promising ZSB. The inoculation of the ZO11 and ZO14 isolates in the corn plant led to an increase in Zn absorption in the roots by 179.7% and 62.37% and by 155.1% and 110.6% in the shoot part of the corn plant respectively, compared to the negative control treatment. The results of molecular identification showed that two isolates belonged to *Acinetobacter calcoaceticus* and *Agromyces italicus*.

Fungi as ZSF

One of the important parts of soil microbes is fungi and generally make up more soil biomass than bacteria, depending on soil depth and nutrient conditions. Fungi reportedly have a greater ability to dissolve insoluble metal compounds containing phosphorus, potassium, and zinc. Organic acids (e.g., gluconic acid, malonic acid, lactic acid) production by different ZSM is a common mechanism for making insoluble compounds like Zn. The produced organic acids lower the pH of the medium and the immobile compounds get soluble in lower pH. Raulin (1869) first documented the importance of Zinc for growth and development of fungi while working with *Aspergillus niger*. Fungi belonging to the genus *Aspergillus*, *Penicillium*, *Trichoderma*, *Phlebopus*, *Pisolithus*, and *Suillus* are reported to have Zn solubilizing potential in different *in vitro* studies for several Zn compounds like ZnO, $Zn_3(PO_4)_2$, and $ZnCO_3$ (Kumla et al., 2014; Nath et al., 2015; Sallam et al., 2021; Sayer et al., 1995; Zhang et al., 2021).

Among all other fungi, arbuscular mycorrhizal fungi (AMF) are a special group of fungi which establish mutualistic association with host plants and facilitate their hosts with additional supply of water and nutrients. Members of AMF enhance the nutrient and water scavenging area for the host plant's roots. Like other fungi, AMF can reduce the rhizosphere pH and produce siderophores for solubilizing immobile Zn. In addition, they can transport plant essential nutrients through the external mycelium along with helping other modes of metal solubilization (Suganya et al., 2021). AMF can induce soil enzymatic activity and regulate the glycoprotein secretion which may help in solubilizing Zn (Wamberg et al., 2003).

Mechanism of Zn Solubilization

By using different mechanisms such as proton secretion, production of organic and inorganic acids, siderophore production, etc., ZSM lead to an increase in the bioavailability of elements such as Zn, Fe, P, K, etc. (Khoshru et al., 2020b). Among the factors affecting the solubilization of Zn, pH has the most important effect on increasing the solubility of Zn. Thus, soil with higher pH has less Zn than soil with lower pH (Sadeghzadeh, 2012). The production of various organic ligands by ZSM leads to the formation of an organic-Zn complex and increases its availability for microbes and plants. It has been reported that treatment with humic acid increases the adsorption capacity of soil Zn by 73-95% (Piri et al., 2019). It has also been reported that citric acid increased plants Zn uptake by 43% (Ke et al., 2020). It has been reported that changes in soil pH may also improve the effectiveness of organic ligands in increasing Zn plant availability (Moreno-Lora and Delgado, 2020; Salinitro et al., 2020). Figure 3 shows some of the mechanisms reported for ZSB in promoting plant growth.

There are various reports about increasing the Zn content of plants by PGPR, for example, *Bacillus aryabhattai* (Ramesh et al., 2014), *Bacillus* sp. and *Azospirillum* sp. (Hussein et al., 2015). Some bacterial strains have performed Zn solubilization under in vitro conditions, such as *Bacillus* sp., *Pseudomonas striata*, *P. fluorescence* (Abaid-Ullah et al., 2015), *Gluconacetobacter diazotrophicus* (Saravanan et al., 2007b), *Serratia liquefaciens*, *S. marcescens*, *Bacillus thuringiensis* (Abaid-Ullah et al., 2015) and *P. aeruginosa* (Fasim et al., 2002). ZSB has led to the enhancement of Zn absorption in various plants such as maize, soybean and wheat (Khande et al., 2017). Vaid et al. (2014) reported that inoculation of ZSB to rice plants resulted in a 42.7% increase in Zn uptake.

It has been reported that ZSM can indirectly contribute to the absorption of soil mineral elements such as zinc by affecting the root architecture. Verma et al. (2021) reported that ZSB by producing phytohormones such as auxin can modify the root architecture e.g., increasing the length of the root and the density of its lateral hairs, and increasing the active surface of the root. This root changes help to the plant to discover more soil volume for water and nutrients and cause an overall improvement in plant growth. There are reports of changes in plant root morphology and architecture under conditions of zinc deficiency, for example, Genc et al. (2007) reported that zinc deficiency in barley led to an increase in root length. It has also been reported that zinc deficiency increased the length of the primary root system of rice (Zeng et al.,

2019). Zinc deficiency in *Arabidopsis* led to an increase in the number and length of lateral roots, but the length of the primary root decreased (Jain et al., 2013).

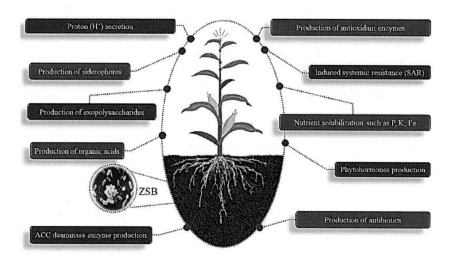

Figure 3. Some of the plant growth promoting mechanisms by ZSB.

Role of Zn-Solubilizing in Sustainable Agriculture

Microorganisms of soil are the important constituents responsible for maintaining soil health and mineralization of soil nutrients that can be used for cultivation. Many microbial species are used as biological fertilizers or biological control agents to improve soil health and crop yields, thereby increasing farmers' income and living standards. Zinc is such an important micronutrient that is a prerequisite for different plants from their earlier stage of growth till the harvest. However, Zn deficiency is still prevailing in different agricultural soils worldwide. To supply ample amounts of Zn to plants, farmers fertilize their fields with Zn fertilizers almost every season, which increases the production cost. Most of the applied Zn gets fixed in soil in a short period of time, which renders plants devoid of sufficient Zn supply. Besides, using synthetic fertilizers also enhances the concern for environmental pollution. Therefore, sustainable management of Zn fertilizers using ZSM can be a potential tool for efficient use of Zn fertilizers without causing much damage to the environment.

Several studies have been conducted to find out the beneficial use of ZSM for crop production. Rion et al., (2022) reported that Zn solubilizing rhizobacterial isolates can augment the early seedling growth of rice. Rice plant growth depends on the inoculated ZSB, and the sources and rate of zinc (Othman et al., 2017). Bacterial inoculations escalated total Zn uptake in rice along with substantial improvement in vegetative growth parameters and number of productive tillers compared to uninoculated control (Ash et al., 2020; Shakeel et al., 2015; Vaid et al., 2014). Application of ZSM improved grain Zn uptake, root and shoot length, root and shoot fresh weight and dry weight in maize (Ayyar and Appavoo, 2016; Bashir et al., 2021; Hussain et al., 2015; Mumtaz et al., 2017). The growth and nutrient quality of another cereal crop wheat also tend to improve when fertilized with ZSM. In a pot experiment, wheat cultivated with *Pseudomonas* sp. resulted in 31% more Zn in grain than the untreated control (Joshi et al., 2013). When wheat seedlings were inoculated with five microbes (*P. fragi, P. dispersa, P. agglomerans, E. cloacae,* and *Rhizobium* sp.) individually, they boosted the root and shoot lengths, root and shoot dry weights, and grain Zn content (Kamran et al., 2017; Khande et al., 2017; Yadav et al., 2020). Other crops like soybean, chickpea, tomato, cucumber, guava etc. were also benefited with enhanced growth and nutrition when cultivated with different ZSM (Chen et al., 2017; Khande et al., 2017; Panneerselvam et al., 2012; Ramesh et al., 2014; Shahid et al., 2020; Sharma, 2012; Yasmin et al., 2021). Application of ZSM can improve the crop quality and maintain the sufficient supply of Zn in soil solution for plant uptake. Therefore, employing ZSM for field use may be a sustainable approach for ensuring proper Zn supply in plants and to mitigate existing micronutrient deficiency in underdeveloped countries.

Conclusion

Although Zn is an important micronutrient for the production of agricultural products, the deficiency of this element is common in arid and semi-arid regions. Reasons such as calcareous soil, high pH, high percentage of calcium carbonate, lack of organic matter and high consumption of phosphorus fertilizers have caused that the amount of Zn be less than the amount necessary to ensure the optimal growth of most agricultural products. One of the main solutions to compensate for the lack of this nutrient is the use of chemical fertilizers, which are not very effective and efficient in calcareous and alkaline soils. Therefore, agricultural systems cannot be stable without changing the situation and need new approaches. Research conducted in different regions of the world has shown that using the potential of microorganisms, especially soil rhizobacteria, can lead to the supply of Zn to plants through an eco-friendly approach. According to the contents presented in this chapter, it seems that in order to achieve sustainable agriculture in the near future, it is necessary to use fewer chemical inputs along with more reliance on such beneficial microorganisms to solubilize and mobilize soil nutrients and promote plant growth and development. However, there is an urgent need to carry out more studies related to improving screening techniques, isolation and characterization of ZSB. More research is necessary in future to explore the potentiality of ZSB or ZSF in soil fertility.

References

Abaid-Ullah M, Hassan MN, Jamil M, Brader G, Shah MKN, Sessitsch A, et al. Plant growth promoting rhizobacteria: an alternate way to improve yield and quality of wheat (*Triticum aestivum*). *Int J Agric Biol*. (2015);17:51-60.

Ahmad I, Ahmad M, Hussain A, Jamil M. Integrated use of phosphate-solubilizing Bacillus subtilis strain IA6 and zinc-solubilizing Bacillus sp. strain IA16: a promising approach for improving cotton growth. *Folia Microbiol* (Praha). (2021);66(1):115-125. doi: 10.1007/s12223-020-00831-3.

Ahmad, S, Mfarrej, MFB, El-Esawi, MA, Waseem, M, Alatawi, A, Nafees, M, Saleem, MH, Rizwan, M, Yasmeen, T, Anayat, A, & Ali, S. Chromium-resistant Staphylococcus aureus alleviates chromium toxicity by developing synergistic relationships with zinc oxide nanoparticles in wheat. *Ecotoxicol Environ Saf*. (2022);230:113142. doi: 10.1016/j.ecoenv.2021.113142.

Al Jabri H, Saleem MH, Rizwan M, Hussain I, Usman K, Alsafran M. Zinc oxide nanoparticles andtheir biosynthesis: Overview. *Life*. (2022);12(4):594. doi: 10.3390/life12040594.

Alloway BJ. Soil factors associated with zinc deficiency in crops and humans. *Environ Geochem Health*. (2009);31(5):537-548.

Alloway BJ. *Zinc in Soil and Crop Nutrition*. Brussels, Belgium: International Zinc Association; (2004).

Alloway BJ. *Zinc in Soils and Plant Nutrition*. Belgium: International Zinc Association; (2008).

Altomare C, Norvell WA, Björkman T, Harman GE. Solubilization of Phosphates and Micronutrients by the Plant-Growth-Promoting and Biocontrol Fungus Trichoderma harzianum Rifai 1295-22. *Appl Environ Microbiol*. (1999);65(7):2926-2933. doi: 10.1128/AEM.65.7.2926-2933.1999.

Zinc Solubilizing Bacteria 51

Ash M, Yadav J, Yadav JS. Yield attributes of rice (Oryza sativa) as affected by integrated use of zinc oxide and zinc solubilizers. *Indian J Agric Sci.* (2020);90:2180-2184. doi: 10.56093/ijas.v90i11.108591.

Ayyar S, Appavoo S. Effect of Graded Levels of Zn in Combination with or without Microbial Inoculation on Zn Transformation inSoil, Yield and Nutrient Uptake by Maize for Black *Soil. Environ Ecol Res.* (2016);35:172-176.

Baig KS, Arshad M, Shaharoona B, Khalid A, Ahmed I. Comparative effectiveness of Bacillus spp. possessing either dual or single growth-promoting traits for improving phosphorus uptake, growth and yield of wheat (*Triticum aestivum* L.). *Ann Microbiol.* (2012);62:1109-1119. doi: 10.1007/s13213-011-0352-0.

Balakrishnan N, Subramanian KS. Mycorrhizal (Rhizophagus intraradices) Symbiosis and Fe and Zn Availability in Calcareous Soil. *Commun Soil Sci Plant Anal.* (2016);47(10):1357-1371. doi: 10.1080/00103624.2016.1178278.

Bankaji I, Pérez-Clemente R, Caçador I, Sleimi N. Accumulation potential of atriplex halimus to zinc and lead combined with NaCl: Effects on physiological parameters and antioxidant enzymes activities. *South Afr J Bot.* (2019);123:51-61. doi: 10.1016/j.sajb.2019.02.011.

Barber SA. *Soil Nutrient Bioavailability: A Mechanistic Approach.* John Wiley & Sons; 1995.

Bashir S, Basit A, Abbas RN, Naeem S., Bashir Sa, Ahmed N, Ahmed MS, Ilyas MZ, Aslam Z, Alotaibi SS, El-Shehawi AM, Li Y. Combined application of zinc-lysine chelate and zinc-solubilizing bacteria improves yield and grain biofortification of maize (*Zea mays* L.). *PLoS One.* (2021);16(7):e0254647. doi: 10.1371/journal. pone.0254647.

Batool S, Asghar HN, Shehzad MA, Yasin S, Sohaib M, Nawaz F, Akhtar G, Mubeen K, Zahir ZA, Uzair M. Zinc-Solubilizing Bacteria-Mediated Enzymatic and Physiological Regulations Confer Zinc Biofortification in Chickpea (*Cicer arietinum* L.). *J Soil Sci Plant Nutr.* (2021);21:2456-2471. doi: 10.1007/s42729-021-00537-6.

Berendsen RL, Pieterse CM, Bakker PA. The rhizosphere microbiome and plant health. *Trends Plant Sci.* (2012);17(8):478-486. doi: 10.1016/j.tplants.2012.04.001.

Beura K, Singh M, Pradhan AK, Rakshit R, Lal M. Evaluation of Arbuscular Mycorrhiza Fungi Species for Their Efficiency Towards Nutrient Acquisition in Rhizospheric Soil of Maize. *Int J Bio-Sci Bio-Technol.* (2016);7:130-135. doi: 10.23910/IJBSM/ 2016.7.1.1505b.

Bhantana, P, Rana, MS, Sun, X, Moussa, MG, Saleem, MH, Syaifudin, M, Shah, A, Poudel, A, Pun, AB, Bhat, MA, Mandal, DL, Shah, S, Zhihao, D, Tan, Q, & Hu, CX. Arbuscular mycorrhizal fungi and its major role in plant growth, zinc nutrition, phosphorous regulation and phytoremediation. *Symbiosis.* (2021);84(1):1-19. doi: 10.1007/s13199-021-00756-6.

Bhatt K, Maheshwari DK. Decoding multifarious role of cow dung bacteria in mobilization of zinc fractions along with growth promotion of *C. annuum* L. *Sci Rep.* (2019);9:14232. doi: 10.1038/s41598-019-50788-8.

Bonaventura P, Benedetti G, Albarède F, Miossec P. Zinc and its role in immunity and inflammation. *Autoimmun Rev.* (2015);14:277-285. doi: 10.1016/j.autrev.2014. 11.008.

Cakmak I, Pfeiffer WH, McClafferty B. Biofortification of durum wheat with zinc and iron. *Cereal Chem.* (2010);87:10-20. doi: 10.1094/CCHEM-87-1-0010.

Cakmak I, Yilmaz A, Kalayci M, Ekiz H, Torun B, Braun HJ. Zinc deficiency as a critical problem in wheat production in Central Anatolia. *Plant Soil.* (1996);180:165-172. doi: 10.1007/BF00015288.

Cakmak I. Zinc deficiency in wheat in Turkey. In: Alloway BJ, editor. Micronutrient deficiencies in global crop production. *Springer Netherlands*; (2008). p. 181-200. doi: 10.1007/978-1-4020-6860-7_11.

Castillo-González J, Ojeda-Barrios D, Hernández-Rodríguez A, González-Franco AC, Robles-HernándezL, López-Ochoa GR. Zinc metalloenzymes in plants. *Interciencia.* (2018);43:242-248. doi: 10.378-1844/14/07/468-08.

Catlett KM, Heil DM, Lindsay WL, Ebinger MH. Soil chemical properties controlling Zn2+ activity in 18 Colorado soils. *Soil Sci Soc Am J.* (2002);66:1182-1189. doi: 10.2136/sssaj2002.1182.

Chahal DS, Sharma BD, Singh PK. Distribution of forms of zinc and their association with soil properties and uptake in different soil orders in semi-arid soils of Punjab. India. *Commun Soil Sci Plan.* (2005);36:2857-2874. doi: 10.1080/00103620500306031.

Chakraborty S, Mishra AK. Mitigation of zinc toxicity through differential strategies in two species of the cyanobacterium anabaena isolated from zinc polluted paddy field. *Environ Pollut.* (2020);263:114375. doi: 10.1016/j.envpol.2020.114375.

Chattha, MU, Hassan, MU, Khan, I, Chattha, MB, Mahmood, A, Chattha, MU, Nawaz, M, Subhani, MN, Kharal, M, & Khan, S. Biofortification of wheat cultivars to combat zinc deficiency. *Front Plant Sci.* (2017);8:281. doi: 10.3389/fpls.2017.00281.

Chen S, Zhao H, Zou C, Li Y, Chen Y, Wang Z, Jiang Y, Liu A, Zhao P, Wang M, Ahammed GJ. Combined Inoculation with Multiple Arbuscular Mycorrhizal Fungi Improves Growth, Nutrient Uptake and Photosynthesis in Cucumber Seedlings. *Front Microbiol.* (2017);8:2516. doi: 10.3389/fmicb.2017.02516.

Chesworth W. Geochemistry of micronutrients. In: Mortvedt JJ, Cox FR, Shuman LM, Welch RM, editors. Micronutrients in agriculture. *Soil Science Society of America*; (1991). p. 1-30. doi: 10.2136/sssaspecpub53.c1.

Dinesh R, Anandaraj M, Kumar A, Bini YK, Subila KP, Aravind R. Isolation, characterization, and evaluation of multi-trait plant growth promoting rhizobacteria for their growth promoting and disease suppressing effects on ginger. *Microbiol Res.* (2015);173:34-43. doi: 10.1016/j.micres.2015.01.002.

Ebrahimi M, Safari Sinegani AA, Sarikhani MR, Aliasgharzad N. Assessment of Soluble and Biomass K in Culture Medium Is a Reliable Tool for Estimation of K Releasing Efficiency of Bacteria. *Geomicrobiol J.* (2019);36(10):873-880. doi: 10.1080/01490451.2019.1641771.

Ebrahimi M, Sarikhani MR, Safari Sinegani AA, Aliasgharzad N. Inoculation effects of isolated plant growth promoting bacteria on wheat yield and grain N content. *J Plant Nutr.* (2023);46(7):1407-1420. doi: 10.1080/01904167.2022.2037825.

Eshaghi E, Nosrati R, Owlia P, Malboobi MA, Ghaseminejad P, Ganjali MR. Zinc solubilization characteristics of efficient siderophore-producing soil bacteria. *Iran J Microbiol.* (2019);11:419-430.

Fageria NK. Dry matter yield and nutrient uptake by lowland rice at different growth stages. *J Plant Nutr*. (2004);27(6):947-958. doi: 10.1081/PLN-120037444.

Fageria NK. Influence of micronutrients on dry matter yield and interaction with other nutrients in annual crops. *Pesquisa Agropecuária Brasileira*. (2002);37:1765-1772. doi: 10.1590/S0100-204X2002001200012.

Fahsi N, Mahdi I, Mesfioui A, Biskri L, Allaoui A. Plant growth-promoting rhizobacteria isolated from the jujube (*Ziziphus lotus*) plant enhance wheat growth, zn uptake, and heavy metal tolerance. *Agriculture*. (2021);11:316. doi: 10.3390/agriculture11040316.

Fasim F, Ahmed N, Parsons R, Gadd GM. Solubilization of zinc salts by a bacterium isolated from the air environment of a tannery. *FEMS Microbiol Lett*. (2002);213:1-6. doi: 10.1111/j.1574-6968.2002.tb11277.x.

Gandhi A, Muralidharan G. Assessment of zinc solubilizing potentiality of Acinetobacter sp. isolated from rice rhizosphere. *European Journal of Soil Biology*. (2016);76:1-8. doi: 10.1016/j.ejsobi.2016.06.006.

Genc Y, Huang CY, Langridge P. A study of the role of root morphological traits in growth of barley in zinc-deficient soil. *J Exp Bot*. (2007);58(11):2775-2784. doi: 10.1093/jxb/erm142.

Ghavami N, Alikhani HA, Pourbabaee AA, Besharati H. Study the Effects of Siderophore-Producing Bacteria on Zinc and Phosphorous Nutrition of Canola and Maize Plants. *Communications in Soil Science and Plant Analysis*. (2016);47:1517-1527. doi: 10.1080/00103624.2016.1194991.

Giasson P, Jaouich A, Gagné S, Moutoglis P. Arbuscular mycorrhizal fungi involvement in zinc and cadmium speciation change and phytoaccumulation. *Remediation Journal*. (2005);15:75-81. doi: 10.1002/rem.20044.

Gong, B, He, E, Qiu, H, Van Gestel, CAM, Romero-Freire, A, Zhao, L, Xu, X, & Cao, X. Interactions of arsenic, copper, and zinc in soil-plant system: Partition, uptake and phytotoxicity. *Sci Total Environ*. (2020);745:140926. doi: 10.1016/j.scitotenv.2020.140926.

Gontia-Mishra I, Sapre S, Tiwari S. Zinc solubilizing bacteria from the rhizosphere of rice as prospective modulator of zinc biofortification in rice. *Rhizosphere*. (2017);3:185-190. doi: 10.1016/j.rhisph.2017.04.013.

Gupta S, Thokchom SD, Koul M, Kapoor R. Arbuscular Mycorrhiza mediated mineral biofortification and arsenic toxicity mitigation in *Triticum aestivum* L. *Plant Stress*. (2022);5:100086. doi: 10.1016/j.stress.2022.100086.

Hafeez B, Khanif Y, Saleem M. Role of zinc in plant nutrition: A review. *Am J Exp Agric*. (2013);3(2):374. doi: 10.9734/AJEA/2013/2746.

Hashemnejad F, Barin M, Khezri M, Ghoosta Y, Hammer EC. Isolation and Identification of Insoluble Zinc-Solubilising Bacteria and Evaluation of Their Ability to Solubilise Various Zinc Minerals. *J Soil Sci Plant Nutr*. (2021);21:2501-2509. doi:10.1007/s42729-021-00540-x.

Holloway RE. *Zinc as a subsoil nutrient for cereals*. CIMMYT; (1996).

Hotz C, Brown KH. *Assessment of the risk of zinc deficiency in populations and options for its control*. (2004): S91-S204.

Hussain A, Arshad M, Zahir ZA, Asghar M. Prospects of zinc solubilizing bacteria for enhancing growth of maize. *Pakistan Journal of Agricultural Sciences.* (2015);52(4):128-139.

Intorne AC, de Oliveira MVV, Lima ML, da Silva JF, Olivares FL, de Souza Filho GA. Identification andcharacterization of Gluconacetobacter diazotrophicus mutants defective in the solubilization of phosphorus and zinc. *Arch Microbiol.* (2009);191:477-483. doi:10.1007/s00203-009-0472-0.

Jain A, Sinilal B, Dhandapani G, Meagher RB, Sahi SV. Effects of deficiency and excess of zinc on morphophysiological traits and spatiotemporal regulation of zinc-responsive genes reveal incidence of cross talk between micro-and macronutrients. *Environ Sci Technol.* (2013);47(10):5327-5335. doi:10.1021/es400113y.

Jin CW, Ye YQ, Zheng SJ. An underground tale: contribution of microbial activity to plant iron acquisition via ecological processes. *Annals of Botany.* (2014);113(1):7-18.

Joshi D, Negi G, Vaid S, Sharma A. Enhancement of Wheat Growth and Zn Content in Grains by Zinc Solubilizing Bacteria. *Intern Jour of Agricul, Environ, and Biotech.* (2013);6:363. doi:10.5958/j.2230-732X.6.3.004.

Kabata-Pendias A, Pendias H. *Biogeochemistry of trace elements.* PWN, Warszawa; (1999).

Kabata-Pendias A, Pendias H. *Trace Elements in Soils and Plants.* CRC Press; (2001).

Kabata-Pendias A. *Trace elements in soils and plants.* CRC Press; (2000).

Kamran S, Shahid I, Baig DN, Rizwan M, Malik KA, Mehnaz S. Contribution of Zinc Solubilizing Bacteria in Growth Promotion and Zinc Content of Wheat. *Frontiers in Microbiology.* (2017);8.

Karak T, Singh UK, Das S, Das DK, Kuzyakov Y. Comparative efficacy of ZnSO4 and Zn-EDTA application for fertilization of rice (Oryza sativa L.). *Arch Agron Soil Sci.* (2005);51:253-264. doi: 10.1080/03650340400026701.

Karthika K, Rashmi I, Parvathi M. Biological functions, uptake and transport of essential nutrients in relation to plant growth. In: Ahmad P, Rasool S, Wani MR, eds. Plant Nutrients and Abiotic Stress Tolerance. *Springer*; (2018):1-49.

Kaur H, Garg N. Zinc toxicity in plants: A review. *Planta.* (2021);253(6):129. doi: 10.1007/s00425-021-03642-z.

Kayalvizhi K, Kathiresan K. Microbes from wastewater treated mangrove soil and their heavy metal accumulation and Zn solubilization. *Biocatal Agric Biotechnol.* (2019);22:101379. doi: 10.1016/j.bcab.2019.101379.

Ke X, Zhang FJ, Zhou Y, Zhang HJ, Guo GL, Tian Y. Removal of cd, Pb, zn, Cu in smelter soil by citric acid leaching. *Chemosphere.* (2020);255:126690. doi: 10.1016/j.chemosphere.2020.126690.

Khalifa AYZ, Alsyeeh AM, Almalki MA, Saleh FA. Characterization of the plant growth promoting bacterium, Enterobacter cloacae MSR1, isolated from roots of non-nodulating Medicago sativa. *Saudi J Biol Sci.* (2016);23:79-86. doi: 10.1016/j.sjbs.2015.06.008.

Khan MU, Sessitsch A, Harris M, Fatima K, Imran A, Arslan M. Cr-resistant and endophytic bacteria associated with Prosopis juliflora and their potential as phytoremediation enhancing agents in metal-degraded soils. *Front Plant Sci.* (2015);5:755. doi: 10.3389/fpls.2014.00755.

Khande R, Sharma SK, Ramesh A, Sharma MP. Zinc solubilizing Bacillus strains that modulate growth, yield and zinc biofortification of soybean and wheat. *Rhizosphere*. (2017);4:126-138. doi: 10.1016/j.rhisph.2017.09.002.

Khanghahi MY, Ricciuti P, Allegretta I, Terzano R, Crecchio C. Solubilization of insoluble zinc compounds by zinc solubilizing bacteria (ZSB) and optimization of their growth conditions. *Environ Sci Pollut Res*. (2018);25:25862-25868. doi: 10.1007/s11356-018-2638-2.

Khoshgoftarmanesh AH, Afyuni M, Norouzi M, Ghiasi S, Schulin R. Fractionation and bioavailability of zinc (Zn) in the rhizosphere of two wheat cultivars with different Zn deficiency tolerance. *Geoderma*. (2018);309:1-6. doi: 10.1016/j.geoderma.2017.08.019.

Khoshru, B, Mitra, D, Joshi, K, Adhikari, P, Rion, MSI, Fadiji, AE, Alizadeh, M, Priyadarshini, A, Senapati, A, Sarikhani, MR, Panneerselvam, P, Mohapatra, PKD, Sushkova, S, Minkina, T, & Keswani, C. Decrypting the multi-functional biological activators and inducers of defense responses against biotic stresses in plants. *Heliyon*. (2023);9:e13825. doi: 10.1016/j.heliyon.2023.e13825.

Khoshru, B, Mitra, D, Khoshmanzar, E, Myo, EM, Uniyal, N, Mahakur, B, Mohapatra, PKD, Panneerselvam, P, Boutaj, H, Alizadeh, M, Cely, MVT, Senapati, A, & Rani, A. Current scenario and future prospectsof plant growth-promoting rhizobacteria: An economic valuable resource for the agriculture revival under stressful conditions. *Journal of Plant Nutrition*. (2020);43(20):3062-3092. doi: 10.1080/01904167.2020.1834624.

Khoshru B, Mitra D, Mahakur B, Sarikhani M, Mondal R, Verma D, Kumud Pant K. Role of soil rhizobacteria in utilization of an indispensable micronutrient zinc for plant growth promotion. *Journal of Critical Reviews*. (2020);21:4644-4654. doi: 10.31838/jcr.07.12.658.

Khoshru B, Moharramnejad S, Gharajeh NH, Asgari Lajayer B, Ghorbanpour M. Plant microbiome and its important in stressful agriculture. In: Singh D, Singh H, Prasad SM, Choudhary DK, eds. Plant Microbiome Paradigm. *Springer*; (2020):13-48. doi: 10.1007/978-981-15-3903-2_2.

Khoshru B, Sarikhani MR, Reyhanitabar A, Oustan S, Malboobi MA. Evaluation of the ability of rhizobacterial isolates to solubilize sparingly soluble iron under in-vitro conditions. *Geomicrobiol J*. (2022);39:804-815. doi: 10.1080/01490451.2022.2078447.

Khoshru B, Sarikhani MR, Reyhanitabar A, Oustan S, Malboobi MA. Evaluation of the potential of rhizobacteria in supplying nutrients of Zea mays L.plant with a focus on zinc. *Journal of Soil Science and Plant Nutrition*. (2023);1-14. doi: 10.1007/s42729-023-0061-2.

Khuna S, Suwannarach N, Kumla J, Frisvad JC, Matsui K, Nuangmek W, Lumyong S. Growth Enhancement of Arabidopsis (Arabidopsis thaliana) and Onion (Allium cepa) With Inoculation of Three Newly Identified Mineral-Solubilizing Fungi in the Genus Aspergillus Section Nigri. *Frontiers in Microbiology*. (2021);12:705896. doi: 10.3389/fmicb.2021.705896.

Kucey RMN. Effect of Penicillium bilaji on the solubility and uptake of P and micronutrients from soil by wheat. *Can. J. Soil. Sci.* (1988);68:261–270. doi: 10.4141/cjss88-026.

Kumar A, Dewangan S, Lawate P, Bahadur I, Prajapati S. Zinc-solubilizing bacteria: a boon for sustainable agriculture. In: Singh D, Singh H, Prasad SM, Choudhary DK, eds. *Plant Growth Promoting Rhizobacteria for Sustainable Stress Management:* Volume 1: Rhizobacteria in Abiotic Stress Management. Springer; (2019):139-155.

Kumar, V, Yadav, AN, Verma, P, Sangwan, P, Saxena, A, Kumar, K, & Singh, B. β-propeller phytases: diversity, catalytic attributes, current developments and potential biotechnological applications. *Int. J. Biol. Macromolecules.* (2017);98:595–609. doi: 10.1016/j.ijbiomac.2017.01.134.

Kumla J, Suwannarach N, Bussaban B, Matsui K, Lumyong S. Indole-3-acetic acid production, solubilization of insoluble metal minerals and metal tolerance of some sclerodermatoid fungi collected from northern Thailand. *Ann Microbiol.* (2014);64:707–720. doi: 10.1007/s13213-013-0706-x.

Kushwaha P, Srivastava R, Pandiyan K, Singh A, Chakdar H, Kashyap PL, Bhardwaj AK, Murugan K, Karthikeyan N, Bagul SY, Srivastava AK, Saxena AK. Enhancement in Plant Growth and Zinc Biofortificationof Chickpea (*Cicer arietinum* L.) by Bacillus altitudinis. *J Soil Sci Plant Nutr.* (2021);21:922–935. doi: 10.1007/s42729-021-00411-5.

Lefèvre I, Vogel-Mikuš K, Jeromel L, Vavpetič P, Planchon S, Arčon I, Elteren, JTV, Lepoint, G, Gobert, S, Renaut, J, Pelicon, P, & Lutts, S. Differential cadmium and zinc distribution in relation to their physiological impact in the leaves of the accumulating Zygophyllum fabago L. *Plant Cell Environ.* (2014);37:1299–1320. doi: 10.1111/pce.12234.

Li JX, Yang XE, He ZL, Jilani G, Sun CY, Chen SM. Fractionation of lead in paddy soils and its bioavailability to rice plants. *Geoderma.* (2007);141:174–180. doi: 10.1016/j.geoderma.2007.05.006.

Lindsay WL. Inorganic equilibria affecting micronutrients in soils. In: Mortveldt JJ, Cox FR, Shuman LM, Welch RM, eds. Micronutrients in Agriculture. *Soil Science Society of America*; (1991):90-112.

Lucas JA, García-Cristobal J, Bonilla A, Ramos B, Gutierrez-Manero J. Beneficial rhizobacteria from rice rhizosphere confers high protection against biotic and abiotic stress inducing systemic resistance in rice seedlings. *Plant Physiol Biochem.* (2014);82:44–53. doi: 10.1016/j.plaphy.2014.05.007.

Majumdar S, Chakraborty U. Phosphate solubilizing rhizospheric Pantoea agglomerans Acti-3 promotes growth in jute plants. *World J Agric Sci.* (2015);11:401–410. doi: 10.5829/idosi.wjas.2015.11.6.1893.

Maleki M, Mokhtarnejad L, Mostafaee S. Screening of rhizobacteria for biological control of cucumber root and crown rot caused by Phytophthora drechsleri. *Plant Pathol J.* (2011);27:78–84. doi: 10.5423/PPJ.2011.27.1.078.

Mandal LN, Mandal B. Zinc fractions in soils in relation to zinc nutrition of lowland rice. *Soil Science.* (1986);142:141-148.

Mandal LN, Mandal B. Zinc fractions in soils in relation to zinc nutrition of lowland rice. *Soil Sci.* (1986);142(3):141-148.

Martino E, Perotto S, Parsons R, Gadd GM. Solubilization of insoluble inorganic zinc compounds by ericoid mycorrhizal fungi derived from heavy metal polluted sites. *Soil Biol Biochem.* (2003);35(1):133-141. doi: 10.1016/S0038-0717(02)00247-X.

Mattiello EM, Ruiz HA, Neves JC, Ventrella MC, Araújo WL. Zinc deficiency affects physiological and anatomical characteristics in maize leaves. *J Plant Physiol.* (2015);183:138-143. doi: 10.1016/j.jplph.2015.05.014.

Mitra, D, Saritha, B, Janeeshma, E, Gusain, P, Khoshru, B, Abo Nouh, FA, Rani, A, Olatunbosun, AN, Ruparelia, J, Rabari, A, Mosquera-Sánchez, LP, Mondal, R, Verma, D, Panneerselvam, P, Das Mohapatra, PK, & B. E., G. S. Arbuscular mycorrhizal fungal association boosted the arsenicresistance in crops with special responsiveness to rice plant. *Environ Exp Bot.* (2022);193:104681. doi: 10.1016/j.envexpbot.2021.104681.

Moreno-Lora A, Delgado A. Factors determining Zn availability and uptake by plants in soils developed under Mediterranean climate. *Geoderma.* (2020);376:114509. doi: 10.1016/j.geoderma.2020.114509.

Mumtaz MZ, Ahmad M, Jamil M, Hussain T. Zinc solubilizing *Bacillus* spp. potential candidates for biofortification in maize. *Microbiol Res.* (2017);202:51-60. doi: 10.1016/j.micres.2017.06.001.

Natasha, N, Shahid, M, Bibi, I, Iqbal, J, Khalid, S, Murtaza, B, Bakhat, HF, Farooq, ABU, Amjad, M, Hammad, HM, Niazi, NK, & Arshad, M. Zinc in soil-plant-human system: A data-analysis review. *Sci Total Environ.* (2022);808:152024. doi: 10.1016/j.scitotenv.2021.152024.

Nath R, Sharma GD, Barooah M. Plant Growth Promoting Endophytic Fungi Isolated from Tea (*Camellia sinensis*) Shrubs of Assam, India. *AEER.* (2015);13. doi: 10.15666/aeer/1303_877891.

Noulas C, Tziouvalekas M, Karyotis T. Zinc in soils, water and food crops. *J Trace Elem Med Biol.* (2018);49:252-260.

Ochieno DM. Soil Sterilization Eliminates Beneficial Microbes That Provide Natural Pest Suppression Ecosystem Services Against Radopholus similis and *Fusarium Oxysporum* V5w2 in the Endosphere and Rhizosphere of Tissue Culture Banana Plants. *Frontiers in Sustainable Food Systems.* (2022);6:53. doi: 10.3389/fsufs.2022.825944.

Oteino, N, Lally, RD, Kiwanuka, S, Lloyd, A, Ryan, D, Germaine, KJ, & Dowling, DN. Plant growth promotion induced by phosphate solubilizing endophytic *Pseudomonas* isolates. *Front Microbiol.* (2015);6:745. doi: 10.3389/fmicb.2015.00745.

Othman NMI, Othman R, Saud HM, Megat Wahab PE. Effects of root colonization by zinc-solubilizing bacteria on rice plant (*Oryza sativa* MR219) growth. *Agriculture and Natural Resources.* (2018);51:532-537. doi: 10.1016/j.anres.2018.05.004.

Panneerselvam P, Mohandas S, Saritha B, Upreti KK. Glomus mosseae associated bacteria and their influence on stimulation of mycorrhizal colonization, sporulation, and growth promotion in guava (*Psidium guajava* L.) seedlings. *Biol Agric Hort.* (2012);28:1-14.

Pawar A, Ismail S, Mundhe S, Patil VD. Solubilization of insoluble zinc compounds by different microbial isolates in vitro condition. *Int J Trop Agric.* (2015);33:865-869.

Piri M, Sepehr E, Rengel Z. Citric acid decreased and humic acid increased Zn sorption in soils. *Geoderma.* (2019);341:39-45. doi: 10.1016/j.geoderma.2018.12.027.

Prasad AS. Impact of the discovery of human zinc deficiency on health. *J Trace Elem Med Biol.* (2014);28:357-363. doi: 10.1016/j.jtemb.2014.09.002.

Ramesh A, Sharma SK, Sharma MP, Yadav N, Joshi OP. Inoculation of zinc solubilizing Bacillus aryabhattai strains for improved growth, mobilization and biofortification of zinc in soybean and wheat cultivated in Vertisols of central India. *Appl Soil Ecol.* (2014);73:87-96. doi: 10.1016/j.apsoil.2013.08.009.

Rattan RK, Shukla LM. Influence of different zinc carriers on the utilization of micronutrients by rice. *J Indian Soc Soil Sci.* (1991);39(4):808-810.

Raulin J. Etudes Chimique sur al vegetation (Chemical studies on plants). *Annales des Sciences Naturelles Botanique et Biologie Vegetale.* (1869);11:293-299.

Ravi RK, Valli PPS, Muthukumar T. Physiological characterization of root endophytic Fusarium haematococcum for hydrolytic enzyme production, nutrient solubilization and salinity tolerance. *Biocatal Agric Biotechnol.* (2022);43:102392. doi: 10.1016/j.bcab.2022.102392.

Reetha S, Bhuvaneswari G, Thamizhiniyan P, Mycin TR. Isolation of indole acetic acid (IAA) producing rhizobacteria of Pseudomonas fluorescens and Bacillus subtilis and enhance growth of onion (Allim cepa. L). *Int J Curr Microbiol App Sci.* (2014);3:568-574.

Rion MSI, Rahman A, Khatun MJ, Zakir HM, Rashid MH, Quadir QF.Screening of Zinc Solubilizing Plant Growth Promoting Rhizobacteria (PGPR) as Potential Tool for Biofortification in Rice. *J Exp Agric Int.* (2022);44:132-143. doi: 10.9734/jeai/2022/v44i930858.

Sadeghzadeh B, Rengel Z. Zinc in soils and crop nutrition. In: Hawkesford MJ, Barraclough P, eds. *The Molecular and Physiological Basis of Nutrient Use Efficiency in Crops.* Oxford, UK: John Wiley & Sons, Ltd; (2011): 335-375.

Sadeghzadeh B. A review of zinc nutrition and plant breeding. *J. Soil Sci. Plant Nutr.* 2013; 13(4): 905-927. doi: 10.4067/S0718-95162013005000072.

Salinitro M, van der Ent A, Tognacchini A, Tassoni A. Stress responses and nickel and zinc accumulation in different accessions of Stellaria media (L.) vill. in response to solution pH variation in hydroponic culture. *Plant Physiol. Biochem.* (2020); 148: 133-141. doi: 10.1016/j.plaphy.2020.01.012.

Sallam N, Ali EF, Seleim MAA, Khalil Bagy HMM. Endophytic fungi associated with soybean plants and their antagonistic activity against Rhizoctonia solani. *Egyptian Journal of Biological Pest Control.* (2021); 31: 54. doi: 10.1186/s41938-021-00402-9.

Saravanan VS, Kalaiarasan P, Madhaiyan M, Thangaraju M. Solubilization of insoluble zinc compounds by Gluconacetobacter diazotrophicus and the detrimental action of zinc ion (Zn2+) and zinc chelates on root knot nematode Meloidogyne incognita. *Lett Appl Microbiol.* (2007); 44(3): 235-241.

Saravanan VS, Madhaiyan M, Thangaraju M. Solubilization of zinc compounds by the diazotrophic, plant growth promoting bacterium Gluconacetobacter diazotrophicus. *Chemosphere.* 2007; 66(9): 1794-1798.

Saravanan VS, Subramoniam SR, Raj SA. Assessing in vitro solubilization potential of different zinc solubilizing bacterial (ZSB) isolates. *Braz J Microbiol.* (2004); 35: 121-125.

Sarikhani MR, Aliasgharzad N, Khoshru B. P solubilizing potential of some plant growth promoting bacteria used as ingredient in phosphatic biofertilizers with emphasis on growth promotion of Zeamays L. *Geomicrobiol J.* (2020). doi: 10.1080/01490451. 2019.1700323.

Sarikhani MR, Khoshru B, Greiner R. Isolation and identification of temperature tolerant phosphate solubilizing bacteria as a potential microbial fertilizer. *World J Microbiol Biotechnol.* (2019); 35: 1-10.

Sarikhani MR, Malboobi MA, Aliasgharzad N, Greiner R. Identification of two novel bacterial phosphatase-encoding genes in Pseudomonas putida strain P13. *J Appl Microbiol.* (2019); 127(4): 1113-1124.

Sarikhani MR, Oustan S, Ebrahimi M, Aliasgharzad N. Isolation and identification of potassium-releasing bacteria in soil and assessment of their ability to release potassium for plants. *Eur J Soil Sci.* (2018); 69: 1078-1086.

Sayer JA, Raggett SL, Gadd GM. Solubilization of insoluble metal compounds by soil fungi: development of a screening method for solubilizing ability and metal tolerance. *Mycol Res.* 1995; 99: 987-993. doi: 10.1016/S0953-7562(09)80762-13.

Selvakumar G, Joshi P, Nazim S, Mishra PK, Bisht JK, Gupta HS. Phosphate solubilization and growth promotion by Pseudomonas fragi CS11RH1 (MTCC 8984), a psychrotolerant bacterium isolated from a high altitude Himalayan rhizosphere. *Biologia.* (2009); 64: 239-245. doi: 10.2478/s11756-009-0041-7.

Shahid I, Tariq K, Mehnaz S. Zinc Solubilizing Fluorescent Pseudomonads as Biofertilizer for Tomato (Solanum lycopersicum L.) under Controlled Conditions. *Asian J Plant Sci Res.* (2020); 10: 1-7.

Shaikh S, Saraf M. Biofortification of Triticum aestivum through the inoculation of zinc solubilizing plant growth promoting rhizobacteria in field experiment. *Biocatal Agric Biotechnol* (2017); 9: 120-126. doi: 10.1016/j.bcab.2016.12.008.

Shakeel M, Rais A, Hassan MN, Hafeez FY. Root associated Bacillus sp. improves growth, yield and zinc translocation for basmati rice (Oryza sativa) varieties. *Front Microbiol* (2015); 6: 186-193.

Sharma SK. Characterization of Zinc-Solubilizing Bacillus Isolates and their Potential to Influence Zinc Assimilation in Soybean Seeds. *J Microbiol Biotechnol.* (2012);22:352-359. doi: 10.4014/jmb.1106.05063.

Shuman L, Wang J. Effect of rice variety on zinc, cadmium, iron, and manganese content in rhizosphere and non-rhizosphere soil fractions. *Commun Soil Sci Plant Anal.* (1997);28:23-26. doi: 10.1080/00103629709369769.

Sidhu GPS, Bali AS, Singh HP, Batish DR, Kohli RK. Insights into the tolerance and phytoremediation potential of Coronopus didymus L.(Sm) grown under zinc stress. *Chemosphere.* (2020);244:125350. doi: 10.1016/j.chemosphere.2019.125350.

Singh B, Natesan SKA, Singh BK, Usha K. Improving zinc efficiency of cereals under zinc deficiency. *Curr Sci.* (2005);88:36-44.

Sirohi G, Upadhyay A, Srivastava PS, Srivastava S. PGPR mediated Zinc biofertilization of soil and its impact on growth and productivity of wheat. *J Soil Sci Plant Nutr.* (2015);15:202-216.

Sofo A, Moreira I, Gattullo CE, Martins LL, Mourato M. Antioxidant responses of edible and model plant species subjected to subtoxic zinc concentrations. *J Trace Elem Med Biol.* (2018);49:261-268. doi: 10.1016/j.jtemb.2018.02.010.

Stein AJ, Nestel P, Meenakshi JV, Qaim M, Sachdev HPS, Bhutta ZA. Plant breeding to control zinc deficiency in India: how cost-effective is biofortification? *Public Health Nutr.* (2007);10:492-501.

Suganya A, Saravanan A, Baskar M, Pandiyarajan P, Kavimani R. Agronomic biofortification of maize (Zea mays L.) with zinc by using of graded levels of zinc in combination with zinc solubilizing bacteria and Arbuscular mycorrhizal fungi. *J Plant Nutr.* (2021);44:988-994. doi: 10.1080/01904167.2020.1845383.

Sultan AAYA, Gebreel HM, Youssef HIA. Biofertilizer effect of some zinc dissolving bacteria free and encapsulated on Zea mays growth. *Arch Microbiol.* (2023);205:202. doi: 10.1007/s00203-023-03537-5.

Tan, S, Han, R, Li, P, Yang, G, Li, S, Zhang, P, Wang, W-B, Zhao, W-Z, & Yin, L-P. Over-expression of the MxIRT1 gene increases iron and zinc content in rice seeds. *Transgenic Res.* (2015);24:109-122. doi: 10.1007/s11248-014-9822-z.

Tariq M, Hameed S, Malik KA, Hafeez FY. Plant root associated bacteria for zinc mobilization in rice. *Pak J Bot.* (2007);39:245.

Tavallali V, Rahemi M, Eshghi S, Kholdebarin B, Ramezanian A. Zinc alleviates salt stress and increases antioxidant enzyme activity in the leaves of pistachio (Pistacia vera L. 'Badami') seedlings. *Turk J Agr Forest.* (2010);34:349-359. doi: 10.3906/tar-0905-10.

Vaid SK, Kumar B, Sharma A, Shukla AK, Srivastava PC. Effect of Zn solubilizing bacteria on growth promotion and Zn nutrition of rice. *J Soil Sci Plant Nutr.* (2014);14:889-910. doi: 10.4067/S0718-95162014005000071.

Verma, P, Yadav, AN, Khannam, KS, Panjiar, N, Kumar, S, Saxena, AK, & Suman, A. Assessment of genetic diversity and plant growth promoting attributes of psychrotolerant bacteria allied with wheat (Triticum aestivum) from the northern hills zone of India. *Ann Microbiol.* (2015);65:1885-1899. doi: 10.1007/s13213-014-1027-4.

Verma PK, Verma S, Chakrabarty D, Pandey N. Biotechnological approaches to enhance zinc uptake and utilization efficiency in cereal crops. *J Soil Sci Plant Nutr.* (2021);21:2412-2424. doi: 10.1007/s42729-021-00532-x.

Wamberg C, Christensen S, Jakobsen I, Müller AK, Sørensen SJ. The mycorrhizal fungus (*Glomus intraradices*) affects microbial activity in the rhizosphere of pea plants (Pisum sativum). *Soil Biol Biochem.* (2003);35:1349-1357. doi: 10.1016/S0038-0717(03)00214-1.

Wang, Y, Yang, X, Zhang, X, Dong, L, Zhang, J, Wei, Y, Feng, Y, & Lu, L. Improved plant growth and Zn accumulation in grains of rice (Oryza sativa L.) by inoculation of endophytic microbes isolated from a Zn hyperaccumulator, Sedum alfredii H. *J Agric Food Chem.* (2014);62:1783-1791. doi: 10.1021/jf404152u.

Watts-Williams SJ, Cavagnaro TR. Arbuscular mycorrhizal fungi increase grain zinc concentration and modify the expression of root ZIP transporter genes in a modern barley (Hordeum vulgare) cultivar. *Plant Sci.* (2018);274:163-170. doi: 10.1016/j.plantsci.2018.05.015.

White PJ, Broadley MR. Biofortifying crops with essential mineral elements. *Trends Plant Sci.* (2005);10:586-593. doi: 10.1016/j.tplants.2005.10.001.

Xie, R, Zhao, J, Lu, L, Ge, J, Brown, PH, Wei, S, Wang, R, Qiao, Y, Webb, SM, & Tian, S. Efficient phloem remobilization of Zn protects apple trees during the early stages of Zn deficiency. *Plant Cell Environ.* (2019);42:3167-3181. doi: 10.1111/pce.13621.

Yadav R, Ror P, Rathore P, Ramakrishna W. Bacteria from native soil in combination with arbuscular mycorrhizal fungi augment wheat yield and biofortification. *Plant Physiol Biochem.* (2020);150:222-233. doi: 10.1016/j.plaphy.2020.02.039.

Yasmin R, Hussain S, Rasool MH, Siddique MH, Muzammil S. Isolation, Characterization of Zn Solubilizing Bacterium (*Pseudomonas protegens* RY2) and its Contribution in Growth of Chickpea (Cicer arietinum L) as Deciphered by Improved Growth Parameters and Zn Content. *Dose-Response.* (2021);19:15593258211036791. doi: 10.1177/15593258211036791.

Zeng H, Wu H, Yan F, Yi K, Zhu Y. Molecular regulation of zinc deficiency responses in plants. *J Plant Physiol.* (2021);261:153419. doi: 10.1016/j.jplph.2021.153419.

Zhang K, Tappero R, Ruytinx J, Branco S, Liao H-L. Disentangling the role of ectomycorrhizal fungi in plant nutrient acquisition along a Zn gradient using X-ray imaging. *Sci Total Environ.* (2021);801:149481. doi: 10.1016/j.scitotenv.2021. 149481.

Zhao K, Wu Y. Effects of Zn deficiency and bicarbonate on the growth and photosynthetic characteristics of four plant species. *PLoS One.* (2017);12(1):e0169812. doi: 10.1371/journal.pone.0169812.

Zlobin IE. Current understanding of plant zinc homeostasis regulation mechanisms. *Plant Physiol Biochem.* (2021);162:327-335. doi: 10.1016/j.plaphy.2021.03.003.

Zuo Y, Zhang F. Iron and zinc biofortification strategies in dicot plants by intercropping with gramineous species. A review. *Agron Sustain Dev.* (2009);29:63-71. doi: 10.1051/agro:2008055.

Chapter 3

The Potential of *Burkholderia* sp. in Meeting the Goals of Sustainable Agriculture

Richa Raghuwanshi*
Seema Devi
and Surya Prakash Dube
Department of Botany, Mahila Mahavidyalaya, Banaras Hindu University,
Varanasi, Uttar Pradesh, India

Abstract

Interest in sustainable approaches to biological control of plant pathogens
and plant-growth augmentation has been driven by environmental and
public concerns. *Burkholderia* sp., a diverse and adaptable bacterial
genus is an incredibly versatile Ggram-negative genus with over 120
species. *Burkholderia,* a phylogenetically coherent genus well-
recognized in agricultural applications are often isolated from infected
patients as well. Recent studies have demonstrated that the pH of the soil
has a significant impact on its biogeographic range. *Burkholderia* sp.
exhibiting laccase activity is beneficial for the survival of the strain in
phenol-rich environments. *Burkholderia* is an important bacterial species
that directly promotes plant growth by phosphate solubilization, nitrogen
fixation, phytohormone production and indirectly by producing
antibiotics and siderophores, inhibiting phytopathogens in diverse crops.
It is also well-known for its bioremediation, biopesticidal activities as
well. The genus holds potential in producing multiple antibiotics by
quorum sensing, a cell-density-dependent regulatory mechanism. The
inhibitory metabolites produced by *Burkholderia* including pyrrolnitrin,
phenylpyrroles, altericidin A, altericidin B, altericidin C, bacteriocins,

---

* Corresponding Author's Email: richabhu@yahoo.co.in.

In: Biofertilizers
Editor: Philip L. Bevis
ISBN: 979-8-89113-082-1
© 2023 Nova Science Publishers, Inc.

and a novel lipopeptide are active against diverse organisms, including the plant parasitic nematode, bacteria, and several important soil-borne plant-pathogenic fungi. The goal of sustainable agriculture is to develop efficient, biological systems that don't need high levels of material inputs. Healthy soil is an important component of sustainability and *Burkholderia* sp. holds immense possibilities to enhance and protect the productivity of the soil. The multifaceted roles exhibited by *Burkholderia* sp. in attaining the sustainable agricultural practices are dealt in detail in the present chapter.

Keywords: *Burkholderia,* bioremediation, biopesticide, quorum sensing, secondary metabolites production, biocontrol agent, plant growth promotion, phytopathogen

1. Introduction

Burkholderia is a sizable group of motile, rod-shaped gram-negative chemoorganotrops with around 120 validly described species inhabiting different environments like soil, water, plants animal, and humans. *Burkholderia cepacia* strains can also endure in the vacuoles of amoebae that are free to move about. The *Burkholderia* genus (Family-Burkholderiaceae; Class- Betaproteobacteria), is a β subdivision of the proteobacteria. Carnations with indications of root rot and wilt were used to isolate the first member of the genus in 1942 (Burkholder, 1942). *Pseudomonas caryophylli* replaced the original name of *Phytomonas caryophylli* for this organism. The genus name was assigned in recognition of the pioneering work of W. Burkholder, who first described the phytopathogen responsible for the bacterial rot of onions in the 1950s (Burkholder, 1950). *Burkholderia* is a genus of rRNA. *Burkholderia* species has been classified into 10 genotypically distinct but phenotypically similar species (genomovars) referring to the *Burkholderia cepacia* complex (Bcc). Based on Polyphasic classification, *B. cepacia* is composed of five distinct genomic species, with a typical genome size of 7-8 Mbp, but the size of the genome varies substantially between the various *Burkholderia* groups. Niche variability of the genus may be due to their high genome plasticity through their large and complex genome sequence, spread over two to four replicons as well as a large number of sequence insertions (Coenye and Vandamme, 2003). The Bcc species has gone through a number of names, including *Pseudomonas cepacia, Pseudomonas multivorans, Pseudomonas kingii,* and *"Eugonic oxidizer."* Seven species were moved from the *Pseudomonas* genus and assigned to the new genus *Burkholderia* depending

The Potential of *Burkholderia* sp. in Meeting the Goals ... 65

on cellular lipid and fatty acid composition, DNA-DNA homology, 16S ribosomal RNA sequences, and behavioral characteristics (Govan et al., 1996). *Burkholderia cepacia* strains were separated into nine genomovars (I-IX). These genomovars were sufficiently genetically diverse to be given the name of new species (Coenye and Vandamme, 2003). The pan-genome is a representation of all the genes found in a species, encompassing the "dispensable" genome and the core genome. *Burkholderia*'s pan-genome is open with a saturation of 86,000-88,000 genes, and its multichromosomal organisation of two or three chromosomes makes it unique. *Burkholderia* pan-genome comprises of 78,782 orthologs, of which 587 genes forming the core genome are highly common. They have recently been divided into three clades based on their varied habitats: Animal and plant pathogens like *Burkholderia glumae*, *Burkholderia pseudomallei*, and *Burkholderia mallei*, as well as the 17 identified species of the *Burkholderia cepacia* complex, which are primarily opportunistic and responsible for human infections in the respiratory tract and cystic fibrosis, make up the majority of Clade I, which retains the genus name *Burkholderia*. Environmental strains make up Clade II, designated as the genus *Caballeronia* (), while environmental bacteria designated as *Paraburkholderia* make up Clade III (Dobritsa and Samadpour, 2016; Sawana et al., 2014). Bcc has the capacity to use a wide range of compounds as carbon sources, and their genome contains insertion sequences that promote genomic plasticity and metabolic adaptability (Miche et al., 2001). *Paraburkholderia* species share traits such as the ability to breakdown aromatic compounds, quorum sensing, and genes for nitrogen fixation and/or nodulation. These traits are useful when interacting with plants, and it has been possible to isolate diazotrophic *Paraburkholderia* species from non-legume plant roots, legume plant nodules, and the rhizosphere (Sheu et al., 2015). *Burkholderia* are beneficial plant-associated species which shows a variety of tasks including colonizing plant surfaces, fix nitrogen, producing siderophores, solubilizing phosphates, metabolizing volatile compounds, destroying pollutants, producing phytohormones, ACC (1-aminocyclopropane-1-carboxylate) deaminase activity and thereby protecting plants from biotic and abiotic stress. Some species of *Burkholderia* are effective against a broad spectrum of phytopathogenic fungi, bacteria, yeast and protozoa (Cain et al., 2000) and can survive under adverse environmental conditions (Duangura et al., 2018). Antibiotic compounds such as phenazine and pyrrolnitrin from *Burkholderia phenazinium*, *Burkholderia pyrrocinia,* and *Burkholderia cepacia* NB-1 play important role in disease suppression. Some features of *B. gladioli* MB39, such as tolerance to low temperatures, ability to assimilate different carbon sources for growth and usage of diverse nutritional elements, make it an ideal candidate for biocontrol. A spectrum of antibiotics produced by *B. gladioli* include gladiolin, toxoflavin, reumycin, enacyloxin and a putative new compound of molecular formula $C_{33}H_{47}Cl_2NO_{13}$ (Depoorter et al., 2021). Notably, *B. gladioli* MB39 have potent activity against *P. digitatum* and *M. phaseolina*. In addition to plant growth promotion and biocontrol properties, some *Burkholderia* species and *Burkholderia cepacia* complex bacteria are also useful for bioremediation, an overview of which has been illustrated in Figure 1. High level of heavy metals in soil inhibits many plants metabolic functions like seed germination, retarding plant growth and slowing root development. Therefore, application of *Burkholderia* sp. can be a good approach to eradicate the heavy metals (U, Zn, Cd) associated problems. The membrane-bound efflux pumps, biosorption/bioaccumulation of the pollutant at or within the bacterial cell membrane, and biomineralization is how *Burkholderia* sp. primarily displays its bioremediating action with the help of enzymes mostly belonging to the oxidoreductases group (Agarwal et al., 2018).

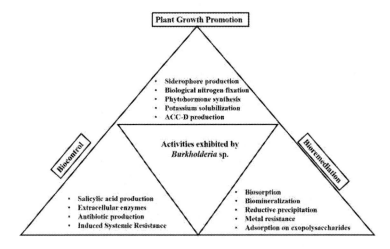

Figure 1. An overview on the mechanisms employed by *Burkholderia* sp. for plant growth promotion, biocontrol and bioremediation.

2. Bacterial Quorum Sensing

Bacterial communication by Quorum sensing (QS) is a key factor in regulating bacterial physiology in the soil. Bacterial QS communication depends on chemical signals that are synthesized and secreted at a higher rate as the bacterial population grows. When the concentration of such signals reaches a certain point, bacterial gene expression can be regulated coordinately, and so can physiological responses to alterations in the environment. Bacterial population in rhizosphere of healthy soil are normally 100-1000 times high than in the bulk soil mostly because bacteria possess metabolic versatility to adapt and utilize root exudates proficiently. The predominant QS signal molecules, produced and detected in proteobacteria, have been shown to be N-acyl-homoserine lactones (AHLs) (Atkinson and Williams, 2009). Few endophytic or rhizospheric *Burkholderia* sp. contributing in food and agricultural areas have been identified as *Burkholderia silvatlantica, Burkholderia tuberum, Burkholderia phymatum, Burkholderia kururiensis, Burkholderia mimosarum, Burkholderia xenovorans,* and *Burkholderia nodosa* (Suárez-Moreno et al., 2012). PGPR isolates are equipped with the BraI/R QS system, which generates 3-O-HSLs, such as 3-O-C14-HSL. Other QS systems, such as XenI2/R2, have also been discovered in some rhizospheric bacteria that generate 3-OH-C8-HSL. Plant colonization is associated with the involvement of BraI/R and XenI2/R2 signaling molecules. The QS systems affect bacterial phenotypes differently, so there may not be a single common target connected to them and they may not be universal regulatory systems (Suárez-Moreno et al., 2010). *Burkholderia phytofirmans* potato strains used to study plant-endophyte interaction was found to produce 3-O-C14, 3-OH-C14, and 3-OH-C12 essentialy. *Burkholderia* species, including *Burkholderia graminis, B. phytofirmans, B. megapolitana,* and *B. bryophila,* have been found to release HSLs when colonizing the tomato rhizosphere (Vandamme et al., 2007). When transgenic plants produced HSLs, the plant growth promoting activities and resistance to salinity got altered in certain species (Barriuso et al., 2008; Trognitz et al., 2008).

3. Growth-Promoting Attributes of *Burkholderia*

Plant growth is the result of the close interrelations between soil nutrient dynamics, climatic conditions and plant microbiome. The endophytic and rhizospheric bacteria have shown great potential in plant growth promotion and development in addition to protection from pathogens and environmental stresses. A range of bacterial strains in this genus are also equipped with biodegrading capabilities. An increasing body of evidence indicates that the various Bcc species are disproportionately distributed in various habitats.

Burkholderia ambifaria, *Burkholderia cenocepacia*, *Burkholderia cepacia*, *Burkholderia vietnamiensis*, and *Burkholderia pyrrocinia* are the most common Bcc species in the plant rhizosphere. *Burkholderia cenocepacia* is better adapted to both the human lung and the plant rhizosphere, due to its higher metabolic diversity and ability to oxidize carbon compounds at high rates (Alisi et al., 2005). Endophytic *Burkholderia* sp. exhibits plant growth promotion through production of phytohormones (IAA, cytokinins, gibberellins), nutrient solubilization, siderophores production, and ACC (1-aminocyclopropane-1-carboxylate) deaminase enzyme activity which overall improves plant growth in terms of root-shoot length, fresh weight, and dry weight, plant nutrient content, chlorophyll content, leaf area, yield, etc. Isolated *Burkholderia* sp. strain FDN2–1 from corn root showed auxin production, potassium as well as phosphorus solubilities (Baghel et al., 2020). Under *in vivo* pot experiments through mechanisms like nitrogen fixation, IAA, siderophore, and 1-aminocyclopropane-1-carboxylate (ACC-D) production, the endophytic bacteria *Burkholderia contaminans* NZ isolated from jute, an important fiber-producing plant, displayed substantial growth promotion activity. The various reports on plant growth promoting attributes exhibited by *Burkholderia* sp. are detailed below.

3.1. Siderophores Production

Iron is one of the most abundant elements on earth and is an important nutrient for bacteria and plant growth. However, in the presence of oxygen and neutral pH, Fe^{2+} is rapidly oxidized to Fe^{3+}, which is not readily available to bacteria. Bacteria have developed ways to scavenge iron with high affinity by producing siderophores, low-molecular-weight chelating molecules that sequester iron from other iron-containing molecules present in the surroundings. Members of the Bcc, *Burkholderia cenocepacia* J2315 can synthesize four different siderophores (pyochelin, ornibactin, cepaciachelin, and cepabactin) for iron chelation and uptake, and can also use ferritin, hemin, and heme as iron sources, allowing Bcc strains to produce hemolysins. (Ratledge and Dover, 2000). Transcriptional analysis of *Burkholderia phytofirmans* inoculated *Arabidopsis* revealed the expression of genes involved in iron storage, siderophore biosynthesis, and transport in shoot tissues (Zhao et al., 2016). Pyochelin, 2-hydroxyphenyl-thiazoline/-oxazoline, and thiazolidine-carboxylate is a siderophore derived from the condensation of two molecules of cysteine with salicylic acid and is produced by some

The Potential of *Burkholderia* sp. in Meeting the Goals ...   69

members of the *Burkholderia cepacia* strains, ATCC 25416 and ATCC 17759, *Burkholderia pseudomallei* and *Burkholderia malle* (Visser et al., 2004). Ornibactin is a linear hydroxamate and α-hydroxy carboxylate siderophore, related in its peptide structure to pyoverdine. It has been isolated from culture supernatants of *Burkholderia ambifaria* and in many other Bcc strains including *Burkholderia cenocepacia* (Barelmann et al., 1996). The genes required for ornibactin biosynthesis and transport have been identified in *Burkholderia cenocepacia* K56-2 and is shown to be negatively regulated by the QS system CepIR (Lewenza and Sokol, 2001). Malleobactin, a hydroxamate siderophore purified from *Burkholderia pseudomallei* K96243, has similarities with ornibactins (Alice et al., 2006). Cepaciachelin (catechols) is another siderophore isolated from a culture supernatant of *Burkholderia cepacia* PHP7 (now *Burkholderia ambifaria*) grown under iron-limiting conditions (Barelmann et al., 1996). Salicylic acid, or 2-hydroxybenzoic acid, was initially identified in *Burkholderia cepacia* isolates and was then called azurechelin (Sokol et al., 2000). Salicylate is a bacterial siderophore that has iron-binding properties and appears to promote iron uptake by the Bcc group promoting growth of plants under iron-limiting conditions (Visser et al., 2004).

3.2. Phytohormones Production

The bacterial production of phytohormones explains the changes in root morphology following *Burkholderia* inoculation. The plant hormones auxins and cytokinins are involved in several stages of plant growth and development, such as cell elongation, cell division, tissue differentiation, and apical dominance. The most important auxin, indole-3 acetic acid, is reported to be produced by several strains of *Burkholderia cepacia* isolated from the rhizosphere, by an endophytic *Burkholderia* isolated from root nodules of *Mimosa pudica*, and by *Burkholderia vietnamiensis* MGK3 isolated from rice root. Inoculation of *Burkholderia vietnamiensis* strain TVV75 increased rice yield up to 22% (Trân Van et al., 2000). Similarly, *Burkholderia phytofirmans* strain PsJN inoculation stimulated grapevine growth and enhanced resistance to cold stress (Ait Barka et al., 2006). *Burkholderia ambifaria* MCI-7 enhanced the growth of *Zea mays* (Ciccillo et al., 2002). In a study, IAA over-producing mutants of endophytic *Burkholderia cepacia* strain RRE25 derived through nitrous acid mutagenesis resulted in better root proliferation and improved nutrient uptake (Zhao et al., 2016). In switchgrass, the inoculated plants were photosynthetically more active and allocated more biomass to above-ground plant parts than to lower-ground parts (Wang et al., 2015). *Burkholderia vietnamiensis*, an endophyte associated with sugarcane had the ability to actively reduce acetylene and produce auxin thereby increasing root and shoot biomass (48.3% and 24.53% respectively), number of tillers (28.5%) as well as yield (5.4%) (Govindarajan et al., 2008).

3.3. Nitrogen Fixation

Table 1. Reports on nitrogen-fixing abilities of *Burkholderia* spp.

Burkholderia sp.	Benevolence Factors	Reference
Free-living and endophytes		
Paraburkholderia phytofirmans	nif, ACC deaminase, EPSs, IAA	Weilharter et al., 2011
Paraburkholderia xenovorans	nif, ACC deaminase	Onofre-Lemus et al., 2009
Paraburkholderia unamae	nif, ACC deaminase	Caballero-Mellado et al., 2007
Paraburkholderia silvatlantica	nif, ACC deaminase	Perin et al., 2006
Paraburkholderia graminis	ACC deaminase	Onofre-Lemus et al., 2009
Paraburkholderia bryophila	Siderophore, antifungal activity, phosphate solubilization	Hsu et al., 2018
Paraburkholderia kururiensis	nif, ACC deaminase, EPSs	Hallack et al., 2010
Caballeronia glathei	nif	Zolg and Ottow, 1975
Paraburkholderia heleia	nif	Aizawa et al., 2010
Paraburkholderia megapolitana	Siderophore, antifungal activity	Vandamme et al., 2007
Paraburkholderia terrae	nif	Yang et al., 2006
Legume nodulating		
Paraburkholderia phenoliruptrix	nod, nif, ACC deaminase	Talbi et al., 2010
Paraburkholderia phymatum	nod, nif, ACC deaminase	Vandamme et al., 2002
Paraburkholderia tuberum	nod, nif, ACC deaminase	Chen et al., 2006
Paraburkholderia mimosarum	nod, nif	Nion and Toyota, 2008
Paraburkholderia caballeronis	nod, nif	Achouak et al., 1999
Paraburkholderia diazotrophica	nod, nif	De Meyer et al., 2014
Paraburkholderia dilworthii	nod, nif	Sheu et al., 2015
Paraburkholderia kirstenboschensis	nod, nif	De Meyer et al., 2013
Paraburkholderia rhynchosiae	nod, nif	Chen et al., 2008
Paraburkholderia sabiae	nod, nif	Weilharter et al., 2011

Abbreviations: nod- Nodulating genes, nif- Nitrogen fixing genes, EPSs- Exopolysaccharides.

Table 2. Biocontrol activities reported for *Burkholderia* sp.

Burkholderia sp.	Microbe inhibited	Inhibitory compound	Reference
B. pyrrocinia, B. cepacia NB-1, B. cepacia B37w	Streptomyces, Proteus vulgaris, Fusarium sambucinum	Pyrrolnitrin [3-chloro-4-(2'-nitro-3'chlorophenyl)-pyrrole]	Chernin et al., 1996
B. pyrrocinia ATCC 39277	Pythium ultimum, R. solani, and Sclerotium rolfsii	Xylocandin Complex	Kang et al., 2004
B. cepacia PCII	P. capsici, F. oxysporum, Pyricularia oryzae, and R. solani	4- quinolinone metabolites (or pseudanes): [2-(2-heptenyl)-3- methyl-4-quinolinone] (HMQ), and 3-methyl-2-(2-nonenyl) - 4-quinolinone (NMQ)	Moon et al., 1996
Burkholderia sp. K481-B101, B. cepacia, B. pseudomallei	fungi and yeasts	Glidobactins	Shoji et al., 1990
B. cepacia CF66	R. solani, Aspergillus flavus, F. oxysporum, and C. albicans	CF661	Quan et al., 2006
Burkholderia strain MP-1	F. oyxsporum, R. solani, and Pythium ultimum	Phenylacetic acid, hydrocinnamic acid, 4-hydroxyphenyl acetic acid, and 4-hydroxy phenylacetate methyl ester	Mao et al., 2006
Burkholderia sp. MSSP	R. solani, P. ultimum, and Botrytis cinerea	2-hydroxymethyl-chroman-4-one	Kang et al., 2004
B. cepacia KB-1	Alternaria kikuchiana	An oligopeptides Altericidins	Kirinuki et al., 1984
B. cepacia SC 11 783	Staphylococci	Cepacins A and B	Parker et al., 1984
B. cepacia D-202	Botrytis cinerea and Penicillium expansum	3-amino-2-piperidinone-containing lipids cepaciamides A and B	Jiao et al., 1996
Burkholderia sp. strain MSSP	Thielaviopsis basicola	Hydrogen cyanide (HCN)	Blumer and Haas, 2000
B. cepacia 5.5B, and B. phenazinium	Gaeumannomyces graminis var. tritici, and R. solani	4,9- dihydroxy phenazine-1,6-dicarboxylic acid dimethyl ester, phenazine-1-carboxylic acid, type of phenazines	Laursen and Nielsen, 2004
B. cepacia RJ3 and ATCC 52796	R. solani	Volatile compound(s)	Kai et al., 2007
B. cepacia	Fungi	Lipopeptide	Jayaswal et al., 1993
B. sacchari	R. solani	Polyhydroxyalkanoates	Brämer et al., 2002

Burkholderia sp.	Microbe inhibited	Inhibitory compound	Reference
B. phenazinium	Fungi	Iodine	Viallard et al., 1998
B. caribensis	Fungi	Exopolysaccharides	Achouak et al., 1999
B. phenazinium, B. pyrrocinia, B. ambifaria and B. cepacia	Fungi	Antibiotic compounds like phenazine and pyrrolnitrin	El-Banna and Winkelmann, 1998
B. vietnamiensis, B. ambifaria, and B. phytofirmans	Fungi	Vitamins and phytohormones	Vandamme et al., 1997
Burkholderia seminalis JRBHU6	Fusarium oxysporum, Aspergillus niger, Microsporum gypseum, Trichophyton mentagrophytes and Trichoderma harzianum	pyrrolo[1,2-a] pyrazine-1,4-dione, hexahydro-3(2-methylpropyl)	Prasad et al., 2021
Burkholderia ambifaria	Pythium, Fusarium, Cylindrocarpon, Botrytis, and Rhizoctonia sp.	NR*	Pedersen et al., 1999
B. ambifaria	Pythium aphanidermatum and Aphanomyces euteiches	NR*	Heungens and Parke, 2000
B. ambifaria	Rhizoctonia solani, Pythium ultimum, Fusarium oxysporum, Sclerotium rolfsii and of the nematode Meloidogyne incognita	NR*	Li et al., 2002
B. ambifaria	Sclerotinia sclerotiorum	NR*	McLoughlin et al., 1992
B. ambifaria	F. oxysporum, Fusarium verticillioides	NR*	Bevivino et al., 2000
B. cenocepacia	Pythium	NR*	Parke and Gurian-Sherman, 2001
P. ginsengiterrae, P. panaciterrae	Fungi		Farh et al., 2015

*NR- Not Reported.

The Potential of *Burkholderia* sp. in Meeting the Goals ... 73

Burkholderia vietnamiensis found in association with roots of rice plants grown in Vietnam was first species of this genus reported to fix N_2. *Burkholderia vietnamiensis* was later observed in the maize and coffee plant rhizospheres and rhizoplanes. Similar N_2-fixing abilities were observed in banana (*Musa* spp.) and pineapple (*Ananas comosus* (L.) Merril) inoculants (Magalhães Cruz et al., 2001; Estrada-De Los Santos et al., 2001). *Burkholderia* species that fix N_2 reported are *Burkholderia kururiensis, Burkholderia brasilensis, Burkholderia tropicalis, Burkholderia tropica, Burkholderia unamae, Burkholderia silvatlantica, Burkholderia tuberum, Burkholderia phymatum, Burkholderia caribensis,* and *Burkholderia silvatlantica* (Caballero-Mellado et al., 2004). The nodulation genes required for nitrogen fixation is acquired by *Burkholderia* through horizontal transfer. The congruence between the phylogenetic trees of the 16S rRNA and nifH genes revealed that *Paraburkholderia*'s common ancestor was a diazotroph, and the ability to fix nitrogen is a critical characteristic of plant-growth-promoting species (Moulin et al., 2001; Suárez-Moreno et al., 2012). The literature well documented with reports confirming the N-fixing and growth promoting abilities of *Burkholderia* sp. as listed in Table 1.

4. Role of *Burkholderia* in Plant Diseases and Stress Management

Burkholderia species are well-known elicitors of host plant resistance against biotic and abiotic stresses. Endophytic bacterial density varies from the rhizosphere to the localization of bacterial endophytes inside the plant. The microscopic studies have revealed that the bacteria first gather on the root surface before colonizing the intercellular gaps of the root tissues. The bacterium secretes cell wall degrading enzymes endoglucanase and endopolygalactouronase to gain entry inside the root tissues. Further, the *Burkholderia* sp. is able to induce a plant defense mechanism as phenolic compound accumulation was observed in various root cortical cells (Rosenblueth and Martínez-Romero, 2006). *Burkholderia* sp. through the secretion of pyrrolnitrin and impairing melanin synthesis is reported to inhibit the fungal disease (Zaman et al., 2021). The biocontrol of *Burkholderia glumae,* causing seedling rot and grain rot of rice was controlled by an engineered strain of *Burkholderia* sp. KJ006 containing the N-acyl-homoserine lactonase gene from *Bacillus thuringiensis.* The engineered strain effectively inhibited the production of quorum-sensing signals and reduced the disease incidence (Cho et al., 2007). Inoculation and colonization of grapevine by *Burkholderia phytofirmans* PsJN effectively controlled the growth of *Botrytis cinerea* through cell wall strengthening, accumulating phenolics in the ectodermal region, mobilization of carbon resources, and forming a biofilm around the fungal mycelium, restricting the pathogen (Miotto-Vilanova et al., 2016). *Burkholderia contaminans* NZ is believed to inhibit the growth of *Macrophomina phaseolina*, a phytopathogen that infects crops globally by secreting catechol, pyrrolnitrin, and other antimicrobial agents that acts against melanin biosynthesis, which may contribute to the observed chromogenic aberration and hyphae growth suppression. The array of inhibitory compounds secreted by various species of *Burkholderia* responsible for pathogen growth inhibition are listed in Table 2.

4.1. Extracellular Enzymes Secretions

The extracellular products of *Burkholderia* are admirably diverse and versatile. Extracellular enzymes with proteolytic, lipolytic, and hemolytic activity are released by *Burkholderia* species. However, only the vital enzymes contributing in biocontrol and pathogenic activity of *Burkholderia* in plant system are discussed in the chapter. Extracellular enzymes secretion is regulated by QS, and secretory machinery is needed to export high-molecular weight exoproducts. ZmpA is a zinc metalloprotease that is important for extracellular protease activity in the Bcc group. The *Burkholderia* species also produce lipases, enzymes that catalyze the hydrolysis and synthesis of esters of glycerol with long-chain fatty acids. Chitosan is a deacetylated derivative of chitin, which has been studied for its beneficial biological activities such as antifungal and antibacterial potentials (Rabea et al., 2003). *Burkholderia gladioli* CHB101 is a plant pathogen responsible for the infection of Gladioli (McCulloch, 1921), and two chitosanases (I and II) have been purified from these cultures. (Shimosaka et al., 2000). Plant pathogens secrete a variety of cell wall-degrading enzymes responsible for the breakdown of polysaccharides that compose cell walls, which helps them invade plant tissues. Pathogenic *Burkholderia cepacia* gen. I-induced maceration of the onion is possibly related to polygalacturonase secretion, which was also detected in *Burkholderia gladioli* and *Burkholderia caryophylli* (Massa et al., 2007). *Burkholderia seminalis* JRBHU6 secreting both chitinase and cellulase enzymes showed strong antifungal activity against the fungal strains *viz;*

The Potential of *Burkholderia* sp. in Meeting the Goals ... 75

Fusarium oxysporum, Aspergillus niger, Microsporum gypseum, Trichophyton mentagrophytes and *Trichoderma harzianum* (Prasad et al., 2021).

5. Bioremediating ~~and~~ Potentials of *Burkholderia*

Soil and water are essential natural resources to support human activities. However, the rapid industrialization and use of dyestuffs, explosives, pesticides and pharmaceuticals have resulted in serious environmental pollution. Bioremediation processes uses microorganism or their products to breakdown or transform pollutants to less toxic or non-toxic elemental and compound forms and this depends on the chemical composition and concentration of contaminants, as well as their availability to microorganisms. Bioremediation is primarily focused on biodegradation and entails the total elimination of organic harmful pollutants into innocuous or naturally occurring components such as carbon dioxide, water, and inorganic compounds that are safe for human, animal, plant, and aquatic life. Decomposing and changing pollutants such as hydrocarbons, oil, heavy metals, pesticides, and dyes to less hazardous forms through enzymatic metabolism is an environment friendly approach leaving no hazardous end products. Many processes and pathways have been identified for the biodegradation of a wide range of organic molecules in the presence and absence of oxygen as illustrated in Figure 2. The majority of the enzymes involved are oxidoreductases.

Microorganisms are considered as to be a cheap, simple, and eco-friendly clean-up method for the detoxification of pollutants. To hold the assurance for detoxification of environmental contaminants, diverse group of microbes have been explored around the globe from different locations and environmental conditions. Microbes reported for bioremediation from oily sludge contaminated soil are *Bacillus megaterium, B. cibi, B. cereus, Pseudomonas aeruginosa* and *Stenotrophomonas acidaminiphila*; for heavy metals are *Shinella, Microbacterium, Micrococcus, Bacillus*; microbes for hydrocarbon bioremediation are *Fomes* sp, *Scopulariopsis brevicaulis, Bacillus, Burkholderia, Enterobacter, Kocuria, Pandoraea* and *Pseudomonas* (Sarkar et al., 2017). Many *Burkholderia* isolates have been isolated for their potent capacities to biodegrade anthropogenic organochemical pollutants. The wide substrate diversity of *Burkholderia* makes them attractive bioremediation agents. *Burkholderia* have been promising not only in the laboratories but also in field trials degrading many xenobiotic compounds including polychlorinated biphenyls (PCBs), trichloroethylene (TCE), organopesticides (2,4-D and 2,4,5-T), polycyclic aromatic hydrocarbons (PAHs) and munitions (Royal demolition explosive [RDX], trinitrotoluenes [TNT], dinitrotoluenes [DNTs]) (Nishino et al., 2000). Few reports on bioremediation of polluted soils by *Burkholderia* are documented below.

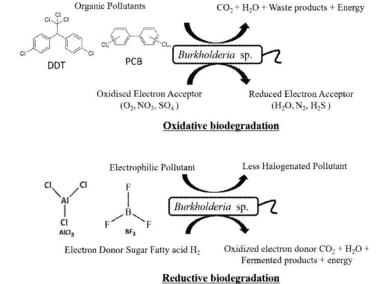

Figure 2. General process of pollutant degradation.

5.1. Mechanism to Cope Up with Uranium Stress

Uranium (U) which is one of the dominant contaminants of the Savannah River Site (SRS) has been microbially converted into less hazardous forms. Strains SRS-25 and SRS-46 of *Burkholderia* spp. isolated from the impacted site showed a rapid depletion of U (Agarwal et al., 2018). Bioremediation involves membrane-bound efflux pumps, metal-resistant genes, detoxifying genes, and biosorption/bioaccumulation of contaminants. In *Burkholderia* sp. metal-resistant genes have been discovered belonging to different categories such as cytochromes, transporters, drug resistance and stress or detoxification, which are able to bioremediate toxins through protein synthesis for cellular survival. Environmental contaminants stimulate gene recruitment, allowing

The Potential of *Burkholderia* sp. in Meeting the Goals ... 77

microbial cells to tolerate and bioremediate toxins (Guo et al., 2010). While the core genes facilitate metabolic tasks, the "foreign" genes, which often exist as orthologous genomic blocks called genomic islands can provide environmental adaptations and genomic flexibility. A *Burkholderia* sp. strain SRS-W-2-2016 isolated from heavy metal enriched soil showed the presence of several genomic islands (GEIs) (Pathak et al., 2017). Upon exposure to uranium some proteins were upregulated in *Burkholderia* sp. SRS-25 and SRS-46 strains. The majority of these proteins were functionally classified into protein production, translocation, damaged DNA repair, and stress response. Bacteria can alter and resist towards U toxicity by three well-known mechanisms viz. reductive precipitation by outer membrane cytochromes, pili, or spores; surface adsorption by exopolysaccharides (EPS) or S-layers; or biomineralization of U via phosphatase enzymes (Das et al., 2016).

5.2. Bioremediation of Zinc Contaminated Environment

Zinc accumulates in environment due to industrial activities like mining and smelting of metal ores, industrial emissions, applications of pesticides and fertilizers, and urban sewage-sludge compost. High level of Zn in soil inhibits many plants metabolic functions like seed germination (Nielsen et al., 2012), retarding plant growth (Michael and Krishnaswamy, 2011), slowing root development (Lingua et al., 2008), inducing foliar chlorosis (Wang et al., 2009), affecting membrane integrity and permeability (Stoyanova and Doncheva, 2002) and interfering with solute uptake, transport, osmotic relations, and the regulation of essential ions (Cherif et al. 2010). Various approaches have been used to remove heavy metal contaminants from soils. Application of growth promoting microbes can be a good approach to eradicate the stress associated with the heavy metals. *Burkholderia* produces secondary metabolites siderophore, IAA and enzyme ACC deaminase, which promote plant growth even under abiotic stress. *Burkholderia pyrrocinia* JK-SH007 are tolerant to heavy metal and synthesize siderophores to solubilize unavailable forms of heavy metal bearing minerals. *Burkholderia cepacia* CS2-1 enhanced the amino acid levels (aspartic acid, threonine, methionine, isoleucine, serine, β-alanine, leucine, tyrosine, glycine, phenylalanine, 3-methylhistidine, valine, γ-aminobutyric acid, alanine, ethanolamine, and proline) in *Brassica rapa* plants grown under both normal and Zn-stressed conditions (Waqas et al., 2016). Among these most of the elevated amino acids were either precursors or intermediates of other stress resistance-related metabolites (Less and Galili, 2008). *Burkholderia cepacia* CS2-1 reduces the Zn-induced oxidative damage of host plants, probably by countering the excess production of reactive oxygen species (ROS). CS2-1 inoculated plant showed significantly lower level of superoxide dismutase SOD, which is a key O_2 scavenger, and generally functions as a first line of defense against abiotic stress-induced injury (Upadhyay et al., 2016). This reduction might be due to the attachment of metals to bacterial surface or from compartmentalization and increased nutrient and biogeochemical cycling in the rhizosphere (Kushwaha et al., 2015). Plants grown in environment with heavy metal contamination accumulate high level of endogenous phytohormone such as ABA that allows them to cope with metal induced stress (Verma et al., 2016). Zn toxicity increases phenylalanine ammonia lyase (PAL) activity, which promotes the synthesis of salicylic acid (SA) from chorismate-derived L-phenylalanine (Luo et al., 2010). *Burkholderia cepacia* CS2-1 eradicate Zn toxicity by enhancing the production of amino acids and reactive oxygen species (ROS) in plants grown under Zn-stress conditions (Waqas et al., 2016). *Burkholderia cepacia* CS2-1 has also been reported to regulate PAL activity, consequently reducing SA synthesis.

5.3. Bioremediation of Cadmium (Cd) Contaminated Environment

Cadmium (Cd) being a highly toxic heavy metal has significant detrimental effects on environment. Cd affects crop growth, inhibits photosynthesis and nucleic acid metabolism even at a very low concentration. Although Cd bioremediation can be driven by using various physico-chemical techniques such as electrolysis, flotation, ion exchange and membrane processed precipitation but these methods are not only expensive but also damages the environment (Nguyen et al., 2013). Microbial bioremediation can be an efficient and eco-friendly alternative. The IncF plasmid of *Burkholderia contaminans* strain ZCC has been reported with multiple genes that shows resistance towards various heavy metals such as As, Cu and Cd. Efflux is the most important Cd resistance mechanism. Efflux of heavy metal pollutants is carried out by an important class of metal transporter called P- type ATPases that use ATP hydrolysis as an energy source to transport ions across cellular membranes. A subfamily of the superfamily P-type ATPases are the PIB-ATPases, important in transport of heavy metals (Rosenzweig and Arguello, 2012). In plasmids heavy metal resistance determinant genes encoding a putative RND-type superfamily responsible for transport of substance from

The Potential of *Burkholderia* sp. in Meeting the Goals ...   79

the periplasm and cytoplasmic membrane across the outer membrane have been identified). ZCC, a gram-negative *Burkholderia* contamination, was isolated from a copper-gold mine. Genomic analysis revealed *B. contaminans*, with 3 chromosomes and 2 plasmids. Presence of three PIB-2-ATPases and 2 Czc-systems and the increased copy number of these determinants due to their location on a plasmid most likely was one of the contributing factors making *Burkholderia contaminans* ZCC more Cd (II) resistant than other strains of *Burkholderia contaminans* as it was able to trigger the synthesis of extracellular polymeric compounds in response to Cd (II) (You et al., 2021).

5.4. Bioremediation of Aromatic Pollutants

A wide range of aromatic substances degraded by *Burkholderia* by using a variety of pathways is well documented (Table 3). Hasan and Jabeen, (2015) reported the examples of microbial degradation of aromatics such as phenol, toluene, aniline, BTX, di-(2-ethylhexyl) phthalate (DEHP), p-cumate, p-cymene, insecticides, and nonanthropogenic aromatic compounds. *Burkholderia* sp. PS3 and *Burkholderia* sp. XTB-5 (Patil and Jena, 2016) has been reported to grow and degrade phenol although the reported mechanism of phenol metabolism by *Burkholderia* remains unclear. Degradation of Ttoluene by *Burkholderia* takes place via three different pathways, all proceeding through either 3-methyl catechol or 4-methylcatechol. *Burkholderia fungorum* converts toluene into a diol and then to 3-methylcatechols, which are subsequently cleaved by catechol-1, 2-dioxygenases to generate muconic acid derivatives (Dobslaw and Engesser, 2015). Recent studies report the metabolic pathways involved in the degradation of aniline, p-hydroxybenzoic, Bbenzene, toluene and xylene (BTX) acid by *Burkholderia* sp. K24 (Lee et al., 2016). Di-(2-ethylhexyl) phthalate (DEHP) is also another human-made aromatic compound that is used as a plasticizer. p-Cymene are formed at significant levels from wood terpenes in the sulfite pulping process (Stromvall and Petersson, 1992). *Burkholderia xenovorans* LB400 was utilized to study not only p-cymene and p-cumate degradation but also the mechanisms to cope with the oxidative stress (Agullo et al., 2017). Multi-omics analyses of *Burkholderia xenovorans* LB400 showed that cmtAb, cymAb, and p-cumate dioxygenases plays important role in the degradation of both compounds. Methyl parathion is one of the most globally used insecticides. A large number of species of genus *Burkholderia* have been reported for degradation of methyl parathion. Out of all these species *Burkholderia zhejiangensis* CEIB S4-3 efficiently degrade and utilize both methyl parathion and its intermediate p-Nitrophenol.

Table 3. Monoaromatic compounds degraded by *Burkholderia* strains

S. No.	Aromatic compounds	*Burkholderia* strain	Reference
1.	Phenol	*Burkholderia* sp. *PS3*	Patil and Jena, 2016
2.	Phenol	*Burkholderia* sp. *XTB-5*	Cheng et al., 2016
3.	Toluene	*Burkholderia* sp. *JS150*	Johnson and Olsen, 1997
4.	Aniline	*Burkholderia* sp. *K24*	Lee et al., 2016
5.	BTX	*Burkholderia* sp. *K24*	Lee et al., 2016
6.	DEHP	*Burkholderia pyrrocinia B1213*	Li et al., 2019
7.	p-Coumaric acid	*Burkholderia glumae BGR1*	Jung et al., 2016
8.	Methyl parathion	*Burkholderia zhejiangensis* CEIBS4-3	Popoca-Ursino et al., 2017
9.	2-Chloro-4-nitrophenol	*Burkholderia* sp. SJ98	Min et al., 2014
10.	3-Methyl-4-nitrophenol	*Burkholderia* sp. SJ98	Min et al., 2016
11.	Naphthalene	*Burkholderia* sp. BC1	Chowdhury et al., 2014
12.	Pyrene	*B. fungorum* T3A13001b	Al-Thukair et al., 2016
13.	Indole	*Burkholderia* sp. IDO3	Ma et al., 2019
14.	Pyridine	*Burkholderia* sp. MAK1	Stankeviciute et al., 2016
15.	Pyrene, phenanthrene, and naphthalene	*B. fungorum* 95 and DBT1	Khoei et al., 2016

Conclusion and Future Prospects

Burkholderia spp. are among the most abundant bacteria in the environment occupying diverse habitats. A better understanding of the specificities and variability in *Burkholderia* individuals may give better insight of their abilities to acclimatize to various environments, in addition to their distinctive interactions which can be exploited to reclaim various natural resources. In the coming years, members of the beneficial *Burkholderia* spp. may be used as alternatives to boost plant development, manage plant diseases and reclaim degraded soils. Molecular tools like CRISPR-Cas9 can be helpful in developing promising strains of *Burkholderia* sp. for various bioremediation and biotechnological purposes. Regaining healthy soil and environment is a prerequisite of any sustainable practice and *Burkholderia* sp. can be a preferable choice in this approach.

Acknowledgments

The authors thankfully acknowledge UGC for providing fellowship and Banaras Hindu University for providing the necessary facilities and grant under the IOE Scheme.

References

Achouak W, Christen R, Barakat M, Martel MH, Heulin T. *Burkholderia caribensis* sp. nov., an exopolysaccharide-producing bacterium isolated from vertisol microaggregates in Martinique. *International Journal of Systematic and Evolutionary Microbiology* (1999) *49*(2):787-794. https://doi.org/10.1099/00207713-49-2-787.

Agarwal M, Pathak A, Rathore RS, Prakash O, Singh R, Jaswal R, Seaman J, Chauhan A. Proteogenomic analysis of *Burkholderia* species strains 25 and 46 isolated from uraniferous soils reveals multiple mechanisms to cope with uranium stress. *Cells* (2018) *7*(12):269. https://doi.org/10.3390/cells7120269.

Agullo L, Romero-Silva MJ, Domenech M, Seeger M. p-Cymene promotes its catabolism through the p-cymene and the p-cumate pathways, activates a stress response and reduces the biofilm formation in *Burkholderia xenovorans* LB400. *PLoS One* (2017) *12*(1): e0169544. https://doi.org/10.1371/journal.pone.0169544.

Ait Barka E, Nowak J, Clément C. Enhancement of chilling resistance of inoculated grapevine plantlets with a plant growth-promoting rhizobacterium, *Burkholderia phytofirmans* strain PsJN. *Applied and Environmental Microbiology* (2006) *72*(11):7246-7252. https://doi.org/10.1128/AEM.01047-06.

Aizawa T, Ve NB, Nakajima M, Sunairi M. *Burkholderia heleia* sp. nov. a nitrogen-fixing bacterium isolated from an aquatic plant, *Eleocharis dulcis,* that grows in highly acidic swamps in actual acid sulfate soil areas of Vietnam. *International Journal of Systematic and Evolutionary microbiology* (2010) 60 (5):1152-1157. https://doi.org/10.1099/ijs.0.015198-0.

Alice AF, López CS, Lowe CA, Ledesma MA, Crosa JH. Genetic and transcriptional analysis of the siderophore malleobactin biosynthesis and transport genes in the human pathogen *Burkholderia pseudomallei* K96243. *Journal of Bacteriology* (2006) 188 (4):1551-1566. https://doi.org/10.1128/JB.188.4.1551-1566.2006.

Alisi C, Lasinio GJ, Dalmastri C, Sprocati A, Tabacchioni S, Bevivino A, Chiarini L. Metabolic profiling of *Burkholderia cenocepacia, Burkholderia ambifaria,* and *Burkholderia pyrrocinia* isolates from maize rhizosphere. *Microbial Ecology* (2005) *50*:385-395. https://doi.org/10.1007/s00248-005-0223-y.

Al-Thukair AA, Malik K. Pyrene metabolism by the novel bacterial strains *Burkholderia fungorum* (T3A13001) and *Caulobacter* sp (T2A12002) isolated from an oil-polluted site in the Arabian Gulf. *International Biodeterioration and Biodegradation* (2016) *110*:32-37. https://doi.org/10.1016/j.ibiod.2016.02.005.

Atkinson S, Williams P. Quorum sensing and social networking in the microbial world. *Journal of the Royal Society Interface* (2009) 6 (40):959-978. https://doi.org/10.1098/rsif.2009.0203.

Baghel V, Thakur JK, Yadav SS, Manna MC, Mandal A, Shirale AO, Sharma P, Sinha NK, Mohanty M, Singh AB, Patra AK. Phosphorus and potassium solubilization from rock minerals by endophytic *Burkholderia* sp. strain FDN2-1 in soil and shift in diversity of bacterial endophytes of corn root tissue with crop growth stage. *Geomicrobiology Journal* (2020) *37*(6):550-563. https://doi.org/10.1080/01490451.2020.1734691.

Barelmann I, Meyer JM, Taraz K, Budzikiewicz H. Cepaciachelin, a new catecholate siderophore from *Burkholderia (Pseudomonas) cepacia. Zeitschrift für Naturforschung C* (1996) *51*(9-10):627-630. https://doi.org/10.1515/znc-1996-9-1004.

Barriuso J, Solano BR, Gutiérrez Mañero FJ. Protection against pathogen and salt stress by four plant growth-promoting rhizobacteria isolated from *Pinus* sp. on *Arabidopsis thaliana. Phytopathology* (2008) *98*(6):666-672. https://doi.org/10.1094/PHYTO-98-6-0666.

Bevivino A, Dalmastri C, Tabacchioni S, Chiarini L. Efficacy of *Burkholderia cepacia* MCI 7 in disease suppression and growth promotion of maize. *Biology and Fertility of Soils* (2000) *31*:225-231. https://doi.org/10.1007/s003740050649.

Blumer C, Haas D. Mechanism, regulation, and ecological role of bacterial cyanide biosynthesis. *Archives of Microbiology* (2000) *173*:170-177. https://doi.org/10.1007/s002039900127.

Bramer CO, Silva LF, Gomez JGC, Priefert H, Steinbuchel A. Identification of the 2-methylcitrate pathway involved in the catabolism of propionate in the polyhydroxyalkanoate-producing strain *Burkholderia sacchari* IPT101T and analysis of a mutant accumulating a copolyester with higher 3-hydroxyvalerate content. *Applied and Environmental Microbiology* (2002) *68*(1):271-279. https://doi.org/10.1128/AEM.68.1.271-279.2002.

Burkholder WH. Three bacterial plant pathogens: *Phytomonas earyophylli* sp. n., *Phytomonas alliicola* sp. n., and *Phytomonas manihotis* (Arthaud-Berthet et Sondar) Viégas. *Phytopathology* (1942) *32*(2):141-149. ISSN: 0031-949X.

Burkholder WH. Sour skin, a bacterial rot of onion bulbs. *Phytopathology* (1950) *40*(1). ISSN: 0031-949X.

Caballero-Mellado J, Martínez-Aguilar L, Paredes-Valdez G, Santos PEDL. *Burkholderia unamae* sp. nov., an N_2-fixing rhizospheric and endophytic species. *International Journal of Systematic and Evolutionary Microbiology* (2004) *54*(4):1165-1172. https://doi.org/10.1099/ijs.0.02951-0.

Caballero-Mellado J, Onofre-Lemus J, Estrada-De Los Santos P, Martínez-Aguilar L. The tomato rhizosphere, an environment rich in nitrogen-fixing *Burkholderia* species with capabilities of interest for agriculture and bioremediation. *Applied and Environmental Microbiology* (2007) *73*(16):5308-5319. https://doi.org/10.1128/AEM.00324-07.

Cain CC, Henry AT, Waldo III RH, Casida Jr LJ, Falkinham III JO. Identification and characteristics of a novel *Burkholderia* strain with broad-spectrum antimicrobial

The Potential of *Burkholderia* sp. in Meeting the Goals ... 83

activity. *Applied and Environmental Microbiology* (2000) *66*(9):4139-4141. https://doi.org/10.1128/AEM.66.9.4139-4141.2000.

Chen WM, James EK, Coenye T, Chou JH, Barrios E, De Faria SM, Elliott GN, Sheu SY, Sprent JI, Vandamme P. *Burkholderia mimosarum* sp. nov., isolated from root nodules of *Mimosa* spp. from Taiwan and South America. *International Journal of Systematic and Evolutionary Microbiology* (2006) *56*(8):1847-1851. https://doi.org/10.1099/ijs.0.64325-0.

Chen WM, de Faria SM, Chou JH, James EK, Elliott GN, Sprent JI, Bontemps C, Young JPW, Vandamme P. *Burkholderia sabiae*, sp. nov., isolated from root nodules of *Mimosa caesalpiniifolia*. *Int J Syst Evol Microbiol* (2008) 58:2174–2179. https://doi.org/10.1099/ijs.0.65816-0.

Cheng Y, Chen Y, Jiang Y, Jiang L, Sun L, Li L, Huang J. Migration of BTEX and Biodegradation in Shallow Underground Water through Fuel Leak Simulation. *Biomed Research International* (2016) 7040872-7040872. https://doi.org/10.1155/2016/7040872.

Cherif J, Derbel N, Nakkach M, Von Bergmann H, Jemal F, Lakhdar ZB. Analysis of in vivo chlorophyll fluorescence spectra to monitor physiological state of tomato plants growing under zinc stress. *Journal of Photochemistry and Photobiology B: Biology* (2010) *101*(3):332-339. https://doi.org/10.1016/j.jphotobiol.2010.08.005.

Chernin L, Brandis A, Ismailov Z, Chet I. Pyrrolnitrin production by an *Enterobacter agglomerans* strain with a broad spectrum of antagonistic activity towards fungal and bacterial phytopathogens. *Current Microbiology* (1996) *32*:208-212. https://doi.org/10.1007/s002849900037.

Cho HS, Park SY, Ryu CM, Kim JF, Kim JG, Park SH. Interference of quorum sensing and virulence of the rice pathogen *Burkholderia glumae* by an engineered endophytic bacterium. *FEMS Microbiology Ecology* (2007) *60*(1):14-23. https://doi.org/10.1111/j.1574-6941.2007.00280.x.

Chowdhury PP, Sarkar J, Basu S, Dutta TK. Metabolism of 2-hydroxy-1-naphthoic acid and naphthalene via gentisic acid by distinctly different sets of enzymes in *Burkholderia* sp. strain BC1. *Microbiology* (2014) 160 (5):892-902. https://doi.org/10.1099/mic.0.077495-0.

Ciccillo F, Fiore A, Bevivino A, Dalmastri C, Tabacchioni S, Chiarini L. Effects of two different application methods of *Burkholderia ambifaria* MCI 7 on plant growth and rhizospheric bacterial diversity. *Environmental Microbiology* (2002) 4 (4):238-245. https://doi.org/10.1046/j.1462-2920.2002.00291.x.

Coenye T, Vandamme P, Govan JR, LiPuma JJ. Taxonomy and identification of the *Burkholderia cepacia* complex. *Journal of Clinical Microbiology* (2001) 39 (10):3427-3436. https://doi.org/10.1128/JCM.39.10.3427-3436.2001.

Coenye T, Vandamme P. Diversity and significance of *Burkholderia* species occupying diverse ecological niches. *Environmental Microbiology* (2003) 5(9):719-729. https://doi.org/10.1046/j.1462-2920.2003.00471.x.

Das S, Dash HR, Chakraborty J. Genetic basis and importance of metal resistant genes in bacteria for bioremediation of contaminated environments with toxic metal pollutants. *Appl. Microbiol. Biotechnol* (2016) 100:2967–2984. https://doi.org/10.1007/s00253-016-7364-4.

De Meyer SE, Cnockaert M, Ardley JK, Trengove RD, Garau G, Howieson JG, Vandamme P. *Burkholderia rhynchosiae* sp. nov., isolated from *Rhynchosia ferulifolia* root nodules. *International Journal of Systematic and Evolutionary Microbiology* (2013) *63*(Pt_11):3944-3949. https://doi.org/10.1099/ijs.0.048751-0.

De Meyer SE, Cnockaert M, Ardley JK, Van Wyk BE, Vandamme PA, Howieson JG. *Burkholderia dilworthii* sp. nov., isolated from *Lebeckia ambigua* root nodules. *International Journal of Systematic and Evolutionary Microbiology* (2014) 64 (4):1090-1095. https://doi.org/10.1099/ijs.0.058602-0.

Depoorter E, De Canck E, Coenye T, Vandamme P. *Burkholderia* bacteria produce multiple potentially novel molecules that inhibit carbapenem-resistant gram-negative bacterial pathogens. *Antibiotics* (2021) *10*(2):147. https://doi.org/10.3390/antibiotics10020147.

Dobritsa AP, Samadpour M. Transfer of eleven species of the genus *Burkholderia* to the genus *Paraburkholderia* and proposal of *Caballeronia* gen. nov. to accommodate twelve species of the genera *Burkholderia* and *Paraburkholderia*. *International Journal of Systematic and Evolutionary Microbiology* (2016) 66 (8):2836-2846. https://doi.org/10.1099/ijsem.0.001065.

Dobslaw D, Engesser KH. Degradation of toluene by ortho cleavage enzymes in *Burkholderia fungorum* FLU 100. *Microbial Biotechnology* (2015) 8 (1):143-154. https://doi.org/10.1111/1751-7915.12147.

Duangurai T, Indrawattana N, Pumirat P. *Burkholderia pseudomallei* adaptation for survival in stressful conditions. *Biomed Research International* (2018) 3039106-3039106. https://doi.org/10.1155/2018/3039106.

El Banna EB, Winkelmann W. Pyrrolnitrin from *Burkholderia cepacia:* antibiotic activity against fungi and novel activities against *Streptomycetes. Journal of Applied Microbiology* (1998) *85*(1):69-78. https://doi.org/10.1046/j.1365-2672.1998.00473.x.

Estrada-De Los Santos P, Bustillos-Cristales R, Caballero-Mellado J. *Burkholderia*, a genus rich in plant-associated nitrogen fixers with wide environmental and geographic distribution. *Applied and Environmental Microbiology* (2001) *67*(6):2790-2798. https://doi.org/10.1128/AEM.67.6.2790-2798.2001.

Farh MEA, Kim YJ, Van An H, Sukweenadhi J, Singh P, Huq MA, Yang DC. *Burkholderia ginsengiterrae* sp. nov. and *Burkholderia panaciterrae* sp. nov., antagonistic bacteria against root rot pathogen *Cylindrocarpon destructans*, isolated from ginseng soil. *Archives of Microbiology* (2015) *197*:439-447. https://doi.org/10.1007/s00203-014-1075-y.

Govan JRW, Hughes JE, Vandamme P. *Burkholderia cepacia*: medical, taxonomic and ecological issues. *Journal of Medical Microbiology* (1996) *45*(6):395-407. https://doi.org/10.1099/00222615-45-6-395.

Govindarajan M, Balandreau J, Kwon SW, Weon HY, Lakshminarasimhan C. Effects of the inoculation of *Burkholderia vietnamensis* and related endophytic diazotrophic bacteria on grain yield of rice. *Microbial Ecology* (2008) *55*:21-37. https://doi.org/10.1007/s00248-007-9247-9.

Guo H, Luo S, Chen L, Xiao X, Xi Q, Wei W, Zeng G, Liu C, Wan Y, Chen J, He Y. Bioremediation of heavy metals by growing hyper accumulaor endophytic bacterium

Bacillus sp. L14. *Bioresource Technology* (2010) *101*(22):8599-8605. https://doi.org/10.1016/j.biortech.2010.06.085.

Hallack LF, Passos DS, Mattos KA, Agrellos OA, Jones C, Mendonça-Previato L, Previato JO, Todeschini AR. Structural elucidation of the repeat unit in highly branched acidic exopolysaccharides produced by nitrogen fixing *Burkholderia*. *Glycobiology* (2010) *20*(3):338-347. https://doi.org/10.1093/glycob/cwp181.

Hasan SA, Jabeen S. Degradation kinetics and pathway of phenol by *Pseudomonas* and *Bacillus* species. *Biotechnology & Biotechnological Equipment* (2015) *29*(1):45-53. https://doi.org/10.1080/13102818.2014.991638.

Heungens K, Parke J. Zoospore homing and infection events: effects of the biocontrol bacterium *Burkholderia cepacia* AMMDR1 on two oomycete pathogens of pea (*Pisum sativum* L.). *Applied and Environmental Microbiology* (2000) *66*(12):5192-5200. https://doi.org/10.1128/AEM.66.12.5192-5200.2000.

Hsu PCL, O'Callaghan M, Condron L, Hurst MR. Use of a gnotobiotic plant assay for assessing root colonization and mineral phosphate solubilization by *Paraburkholderia bryophila* Ha185 in association with perennial ryegrass (*Lolium perenne* L.). *Plant and Soil* (2018) *425*:43-55. https://doi.org/10.1007/s11104-018-3633-6.

Jayaswal RK, Fernandez M, Upadhyay RS, Visintin L, Kurz M, Webb J, Rinehart K. Antagonism of *Pseudomonas cepacia* against phytopathogenic fungi. *Current Microbiology* (1993) *26*:17-22. https://doi.org/10.1007/BF01577237.

Jiao Y, Yoshihara T, Ishikuri S, Uchino H, Ichihara A. Structural identification of cepaciamide A, a novel fungitoxic compound from *Pseudomonas cepacia* D-202. *Tetrahedron Letters* (1996) *37*(7):1039-1042. https://doi.org/10.1016/0040-4039(95)02342-9.

Johnson GR, Olsen RH. Multiple pathways for toluene degradation in *Burkholderia* sp. strain JS150. *Applied and Environmental Microbiology* (1997) *63*(10):4047-4052. https://doi.org/10.1128/aem.63.10.4047-4052.1997.

Juhas M, van der Meer JR, Gaillard M, Harding RM, Hood DW, Crook DW. Genomic islands: Tools of bacterial horizontal gene transfer and evolution. *FEMS Microbiol. Rev.* (2009) 33:376–393. https://doi.org/10.1111/j.1574-6976.2008.00136.x.

Jung DH, Kim EJ, Jung E, Kazlauskas RJ, Choi KY, Kim BG. Production of p-hydroxybenzoic acid from p-coumaric acid by *Burkholderia glumae* BGR1. *Biotechnology and Bioengineering* (2016) *113*(7):1493-1503. https://doi.org/10.1002/bit.25908.

Kai M, Effmert U, Berg G, Piechulla B. Volatiles of bacterial antagonists inhibit mycelial growth of the plant pathogen *Rhizoctonia solani*. *Archives of Microbiology* (2007) *187*:351-360. https://doi.org/10.1007/s00203-006-0199-0.

Kang JG, Shin SY, Kim MJ, Bajpai V, Maheshwari DK, Kang SC. Isolation and anti-fungal activities of 2-hydroxymethyl-chroman-4-one produced by *Burkholderia* sp. MSSP. *The Journal of Antibiotics* (2004) *57*(11):726-731. https://doi.org/10.7164/antibiotics.57.726.

Kirinuki T, Ichiba T, Katayama K. General survey of action site of altericidins on metabolism of *Alternaria kikuchiana* and *Ustilago maydis*. *Journal of Pesticide Science* (1984) *9*(4):601-610. https://doi.org/10.1584/jpestics.9.601.

Khoei NS, Andreolli M, Lampis S, Vallini G, Turner RJ. A comparison of the response of two *Burkholderia fungorum* strains grown as planktonic cells versus biofilm to dibenzothiophene and select polycyclic aromatic hydrocarbons. *Canadian Journal of Microbiology* (2016) *62*(10):851-860. https://doi.org/10.1139/cjm-2016-0160.

Kushwaha A, Rani R, Kumar S, Gautam A. Heavy metal detoxification and tolerance mechanisms in plants: Implications for phytoremediation. *Environmental Reviews* (2015) *24*(1):39-51. https://doi.org/10.1139/er-2015-0010.

Laursen JB, Nielsen J. Phenazine natural products: biosynthesis, synthetic analogues, and biological activity. *Chemical Reviews* (2004) *104*(3):1663-1686. https://doi.org/10.1021/cr020473j.

Lee SY, Kim GH, Yun SH, Choi CW, Yi YS, Kim J, Chung YH, Park EC, Kim SI. Proteogenomic characterization of monocyclic aromatic hydrocarbon degradation pathways in the aniline-degrading bacterium *Burkholderia* sp. K24. *PLoS One* (2016) *11*(4):e0154233. https://doi.org/10.1371/journal.pone.0154233.

Less H, Galili G. Principal transcriptional programs regulating plant amino acid metabolism in response to abiotic stresses. *Plant Physiology* (2008) *147*(1):316-330. https://doi.org/10.1104/pp.108.115733.

Lewenza S, Sokol PA. Regulation of ornibactin biosynthesis and N-acyl-L-homoserine lactone production by CepR in *Burkholderia cepacia*. *Journal of Bacteriology* (2001) *183*(7):2212-2218. https://doi.org/10.1128/JB.183.7.2212-2218.2001.

Li J, Zhang J, Yadav MP, Li X. Biodegradability and biodegradation pathway of di-(2-ethylhexyl) phthalate by *Burkholderia pyrrocinia* B1213. *Chemosphere* (2019) *225*:443-450. https://doi.org/10.1016/j.chemosphere.2019.02.194.

Lingua G, Franchin C, Todeschini V, Castiglione S, Biondi S, Burlando B, Parravicini V, Torrigiani P, Berta G. Arbuscular mycorrhizal fungi differentially affect the response to high zinc concentrations of two registered poplar clones. *Environmental Pollution* (2008) *153*(1):137-147. https://doi.org/10.1016/j.envpol.2007.07.012.

Li W, Roberts DP, Dery PD, Meyer SLF, Lohrke S, Lumsden RD, Hebbar KP. Broad spectrum anti-biotic activity and disease suppression by the potential biocontrol agent *Burkholderia ambifaria* BC-F. *Crop Protection* (2002) *21*(2):129-135. https://doi.org/10.1016/S0261-2194(01)00074-6.

Luo ZB, He XJ, Chen L, Tang L, Gao SHUN, Chen FANG. Effects of zinc on growth and antioxidant responses in *Jatropha curcas* seedlings. *Int J Agric Biol* (2010) *12*(1):119-24. ISSN: 1560-8530.

Magalhães Cruz L, Maltempi de Souza E, Weber OB, Baldani JI, Döbereiner J, de Oliveira Pedrosa F. 16S ribosomal DNA characterization of nitrogen-fixing bacteria isolated from banana (*Musa* spp.) and pineapple (*Ananas comosus* (L.) Merril). *Applied and Environmental Microbiology* (2001) *67*(5):2375-2379. https://doi.org/10.1128/AEM.67.5.2375-2379.2001.

Mao S, Lee SJ, Hwangbo H, Kim YW, Park KH, Cha GS, Park RD, Kim KY. Isolation and characterization of antifungal substances from *Burkholderia* sp. culture broth. *Current Microbiology* (2006) *53*:358-364. https://doi.org/10.1007/s00284-005-0333-2.

Ma Q, Liu Z, Yang B, Dai C, Qu Y. Characterization and functional gene analysis of a newly isolated indole-degrading bacterium *Burkholderia* sp. IDO3. *Journal of*

Hazardous Materials (2019) *367*:144-151. https://doi.org/10.1016/j.jhazmat.2018.12.068.

Massa C, Degrassi G, Devescovi G, Venturi V, Lamba D. Isolation, heterologous expression and characterization of an endo-polygalacturonase produced by the phytopathogen *Burkholderia cepacia*. *Protein Expression and Purification* (2007) *54*(2):300-308. https://doi.org/10.1016/j.pep.2007.03.019.

McCulloch L. A bacterial disease of gladiolus. *Science* (1921) *54*(1388):115-116. https://doi.org/10.1126/science.54.1388.115.

McLoughlin TJ, Quinn JP, Bettermann A, Bookland R. *Pseudomonas cepacia* suppression of sunflower wilt fungus and role of antifungal compounds in controlling the disease. *Applied and Environmental Microbiology* (1992) *58*(5):1760-1763. https://doi.org/10.1128/aem.58.5.1760-1763.1992.

Michael PI, Krishnaswamy M. The effect of zinc stress combined with high irradiance stress on membrane damage and antioxidative response in bean seedlings. *Environmental and Experimental Botany* (2011) *74*:171-177. https://doi.org/10.1016/j.envexpbot.2011.05.016.

Miché L, Faure D, Blot M, Cabanne Giuli E, Balandreau J. Detection and activity of insertion sequences in environmental strains of *Burkholderia*. *Environmental Microbiology* (2001) *3*(12):766-773. https://doi.org/10.1046/j.1462-2920.2001.00251.x.

Min J, Zhang JJ, Zhou NY. The gene cluster for para-nitrophenol catabolism is responsible for 2-chloro-4-nitrophenol degradation in *Burkholderia* sp. strain SJ98. *Applied and environmental microbiology* (2014) *80*(19):6212-6222. https://doi.org/10.1128/AEM.02093-14.

Min J, Lu Y, Hu X, Zhou NY. Biochemical characterization of 3-methyl-4-nitrophenol degradation in *Burkholderia* sp. strain SJ98. *Frontiers in Microbiology* (2016) *7*:791. https://doi.org/10.3389/fmicb.2016.00791.

Miotto-Vilanova L, Jacquard C, Courteaux B, Wortham L, Michel J, Clément C, Barka EA, Sanchez L. *Burkholderia phytofirmans* PsJN confers grapevine resistance against Botrytis cinerea via a direct antimicrobial effect combined with a better resource mobilization. *Frontiers in plant science* (2016) *7*:1236. https://doi.org/10.3389/fpls.2016.01236.

Moon SS, Kang PM, Park KS, Kim CH. Plant growth promoting and fungicidal 4-quinolinones from *Pseudomonas cepacia*. *Phytochemistry* (1996) *42*(2):365-368. https://doi.org/10.1016/0031-9422(95)00897-7.

Moulin L, Munive A, Dreyfus B, Boivin-Masson C. Nodulation of legumes by members of the β-subclass of Proteobacteria. *Nature* (2001) *411*(6840):948-950. https://doi.org/10.1038/35082070.

Nguyen TAH, Ngo HH, Guo WS, Zhang J, Liang S, Yue QY, Li Q, Nguyen TV. Applicability of agricultural waste and by-products for adsorptive removal of heavy metals from wastewater. *Bioresource Technology* (2013) *148*:574-585. https://doi.org/10.1016/j.biortech.2013.08.124.

Nielsen FH. History of zinc in agriculture. *Advances in Nutrition* (2012) *3*(6):783-789. https://doi.org/10.3945/an.112.002881.

88   Richa Raghuwanshi, Seema Devi and Surya Prakash Dube

Nion YA, Toyota K. Suppression of bacterial wilt and *Fusarium* wilt by a *Burkholderia nodosa* strain isolated from Kalimantan soils, Indonesia. *Microbes and Environments* (2008) *23*(2):134-141. https://doi.org/10.1264/jsme2.23.134.

Nishino SF, Paoli GC, Spain JC. Aerobic degradation of dinitrotoluenes and pathway for bacterial degradation of 2, 6-dinitrotoluene. *Applied and Environmental Microbiology* (2000) *66*(5):2139-2147. https://doi.org/10.1128/AEM.66.5.2139-2147.2000.

Onofre-Lemus J, Hernández-Lucas I, Girard L, Caballero-Mellado J. ACC (1-aminocyclopropane-1-carboxylate) deaminase activity, a widespread trait in *Burkholderia* species, and its growth-promoting effect on tomato plants. *Applied and Environmental Microbiology* (2009) *75*(20):6581-6590. https://doi.org/10.1128/AEM.01240-09.

Parke JL, Gurian-Sherman D. Diversity of the *Burkholderia cepacia* complex and implications for risk assessment of biological control strains. *Annual Review of Phytopathology* (2001) *39*(1):225-258. https://doi.org/10.1146/annurev.phyto.39.1.225.

Parker WL, Rathnum ML, Seiner V, Trejo WH, Principe PA, Sykes, RB. Cepacin A and cepacin B, two new antibiotics produced by *Pseudomonas cepacia*. *The Journal of Antibiotics* (1984) *37*(5):431-440. https://doi.org/10.7164/antibiotics.37.431.

Pathak A, Chauhan A, Stothard P, Green S, Maienschein-Cline M, Jaswal R, Seaman J. Genome-centric evaluation of *Burkholderia* sp. strain SRS-W-2-2016 resistant to high concentrations of uranium and nickel isolated from the Savannah River Site (SRS), USA. *Genomics Data* (2017) *12*:62-68. https://doi.org/10.1016/j.gdata.2017.02.011.

Patil SS, Jena HM. Isolation and characterization of phenol degrading bacteria from soil contaminated with paper mill wastewater. *Indian Journal of Biotechnology* (2016) *15*(3): 407-411.

Pedersen EA, Reddy MS, Chakravarty P. Effect of three species of bacteria on damping off, root rot development, and ectomycorrhizal colonization of lodgepole pine and white spruce seedlings. *European Journal of Forest Pathology* (1999) *29*(2):123-134. https://doi.org/10.1046/j.1439-0329.1999.00146.x.

Perin L, Martínez-Aguilar L, Paredes-Valdez G, Baldani JI, Estrada-De Los Santos P, Reis VM, Caballero-Mellado J. *Burkholderia silvatlantica* sp. nov., a diazotrophic bacterium associated with sugar cane and maize. *International Journal of Systematic and Evolutionary Microbiology* (2006) *56*(8):1931-1937. https://doi.org/10.1099/ijs.0.64362-0.

Popoca-Ursino EC, Martínez-Ocampo F, Dantán-González E, Sánchez-Salinas E, Ortiz-Hernández ML. Characterization of methyl parathion degradation by a *Burkholderia zhejiangensis* strain, CEIB S4-3, isolated from agricultural soils. *Biodegradation* (2017) *28*:351-367. https://doi.org/10.1007/s10532-017-9801-1.

Prasad JK, Pandey P, Anand R and Raghuwanshi R. Drought exposed *Burkholderia seminalis* JRBHU6 exhibits antimicrobial potential through Pyrazine-1,4-Dione derivatives targeting multiple bacterial and fungal proteins. Front. Microbiol. (2021). 12:633036. https://doi.org/10.3389/fmicb.2021.633036.

Quan CS, Zheng W, Liu Q, Ohta Y, Fan SD. Isolation and characterization of a novel *Burkholderia cepacia* with strong antifungal activity against *Rhizoctonia solani*.

The Potential of *Burkholderia* sp. in Meeting the Goals ... 89

Applied Microbiology and Biotechnology (2006) *72*:1276-1284. https://doi.org/10.1007/s00253-006-0425-3.

Rabea EI, Badawy MET, Stevens CV, Smagghe G, Steurbaut W. Chitosan as antimicrobial agent: applications and mode of action. *Biomacromolecules* (2003) *4*(6):1457-1465. https://doi.org/10.1021/bm034130m.

Ratledge C, Dover LG. Iron metabolism in pathogenic bacteria. *Annual Reviews in Microbiology* (2000) *54*(1):881-941. https://doi.org/10.1146/annurev.micro.54.1.881.

Rosenblueth M, Martínez-Romero E. Bacterial endophytes and their interactions with hosts. *Molecular Plant-microbe Interactions* (2006) *19*(8):827-837. https://doi.org/10.1094/MPMI-19-0827.

Rosenzweig AC, Argüello JM. Toward a molecular understanding of metal transport by P1B-Type ATPases. In *Current topics in membranes* (2012) 69:113-136 Academic Press. https://doi.org/10.1016/B978-0-12-394390-3.00005-7.

Sarkar P, Roy A, Pal S, Mohapatra B, Kazy SK, Maiti MK, Sar P. Enrichment and characterization of hydrocarbon-degrading bacteria from petroleum refinery waste as potent bioaugmentation agent for in situ bioremediation. *Bioresource Technology* (2017) *242*:15-27. https://doi.org/10.1016/j.biortech.2017.05.010.

Sawana A, Adeolu M, Gupta RS. Molecular signatures and phylogenomic analysis of the genus Burkholderia: proposal for division of this genus into the emended genus *Burkholderia* containing pathogenic organisms and a new genus *Paraburkholderia* gen. nov. harboring environmental species. *Frontiers in Genetics* (2014) *5*:429. https://doi.org/10.3389/fgene.2014.00429.

Sheu SY, Chen MH, Liu WY, Andrews M, James EK, Ardley JK, De Meyer SE, James TK, Howieson JG, Coutinho BG, Chen WM. *Burkholderia dipogonis* sp. nov., isolated from root nodules of *Dipogon lignosus* in New Zealand and Western Australia. *International Journal of Systematic and Evolutionary Microbiology* (2015) *65*(12):4716-4723. https://doi.org/10.1099/ijsem.0.000639.

Shimosaka M, Fukumori Y, Zhang XY, He NJ, Kodaira R, Okazaki M. Molecular cloning and characterization of a chitosanase from the chitosanolytic bacterium *Burkholderia gladioli* strain CHB101. *Applied Microbiology and Biotechnology* (2000) *54*:354-360. https://doi.org/10.1007/s002530000388.

Shoji JI, Hinoo H, Kato T, Hattori T, Hirooka K, Tawara K, Shiratori O, Terui Y. Isolation of cepafungins I, II and III from *Pseudomonas* species. *The Journal of Antibiotics* (1990) *43*(7):783-787. https://doi.org/10.7164/antibiotics.43.783.

Sokol PA, Darling P, Lewenza S, Corbett CR, Kooi CD. Identification of a siderophore receptor required for ferric ornibactin uptake in *Burkholderia cepacia*. *Infection and Immunity* (2000) *68*(12):6554-6560. https://doi.org/10.1128/IAI.68.12.6554-6560.2000.

Stankevičiūtė J, Vaitekūnas J, Petkevičius V, Gasparavičiūtė R, Tauraitė D, Meškys R. Oxyfunctionalization of pyridine derivatives using whole cells of *Burkholderia* sp. MAK1. *Scientific Reports* (2016) *6*(1):39129. https://doi.org/10.1038/srep39129.

Stoyanova Z, Doncheva S. The effect of zinc supply and succinate treatment on plant growth and mineral uptake in pea plant. *Brazilian Journal of Plant Physiology* (2002) *14*:111-116. https://doi.org/10.1590/S1677-04202002000200005.

90 Richa Raghuwanshi, Seema Devi and Surya Prakash Dube

Strömvall AM, Petersson G. Terpenes emitted to air from TMP and sulphite pulp mills. (1992) https://doi.org/10.1515/hfsg.1992.46.2.99.

Suárez-Moreno ZR, Devescovi G, Myers M, Hallack L, Mendonça-Previato L, Caballero-Mellado J, Venturi V. Commonalities and differences in regulation of N-acyl homoserine lactone quorum sensing in the beneficial plant-associated *Burkholderia* species cluster. *Applied and Environmental Microbiology* (2010) 76(13):4302-4317. https://doi.org/10.1128/AEM.03086-09.

Suárez-Moreno ZR, Caballero-Mellado J, Coutinho BG, Mendonça-Previato L, James EK, Venturi V. Common features of environmental and potentially beneficial plant-associated *Burkholderia*. *Microbial Ecology* (2012) *63*:249-266. https://doi.org/10.1007/s00248-011-9929-1.

Talbi C, Delgado MJ, Girard L, Ramírez-Trujillo A, Caballero-Mellado J, Bedmar EJ. *Burkholderia phymatum* strains capable of nodulating *Phaseolus vulgaris* are present in Moroccan soils. *Applied and Environmental Microbiology* (2010) *76*(13):4587-4591. https://doi.org/10.1128/AEM.02886-09.

Trân Van V, Berge O, Ngô Kê S, Balandreau J, Heulin T. Repeated beneficial effects of rice inoculation with a strain of *Burkholderia vietnamiensis* on early and late yield components in low fertility sulphate acid soils of Vietnam. *Plant and Soil* (2000) *218*:273-284. https://doi.org/10.1023/A:1014986916913.

Trognitz F, Scherwinski K, Fekete A, Schmidt S, Eberl L, Rodewald J, Schmid M, Compant S, Hartmann A, Schmitt-Kopplin P, Trognitz B, Sessitsch A. *Interaction between potato and the endophyte Burkholderia phytofirmans* (2008) 63-66. ISBN: 9783902559289.

Upadhyay AK, Singh NK, Singh R, and Rai UN. Ameliration of arsenic toxicity in rice: Comparative effect of inoculation of *Chlorella vulgaris* and *Nannochloropsis* sp. on growth, biochemical changes and arsenic uptake. *Ecotoxicology and Environmental Safety* (2016) 124:68–73. https://doi.org/10.1016/j.ecoenv.2015.10.002.

Vandamme P, Holmes B, Vancanneyt M, Coenye T, Hoste B, Coopman R, Revets H, Lauwers S, Gillis M, Kersters K, Govan JRW. Occurrence of multiple Genomovars of *Burkholderia cepacia* in cystic fibrosis patients and proposal of *Burkholderia multivorans* sp. nov. *International Journal of Systematic Bacteriology* (1997) *47*(4):1188-1200. https://doi.org/10.1099/00207713-47-4-1188.

Vandamme P, Goris J, Chen WM, De Vos P, Willems A. *Burkholderia tuberum* sp. nov. and *Burkholderia phymatum* sp. nov., nodulate the roots of tropical legumes. *Systematic and Applied Microbiology* (2002) *25*(4):507-512. https://doi.org/10.1078/07232020260517634.

Vandamme P, Opelt K, Knöchel N, Berg C, Schönmann S, De Brandt E, Eberl L, Falsen E, Berg G. *Burkholderia bryophila* sp. nov. and *Burkholderia megapolitana* sp. nov., moss-associated species with antifungal and plant-growth-promoting properties. *International Journal of Systematic and Evolutionary Microbiology* (2007) *57*(10):2228-2235. https://doi.org/10.1099/ijs.0.65142-0.

Verma V, Ravindran P, Kumar PP. Plant hormone-mediated regulation of stress responses. *BMC Plant Biology* (2016) *16*: 1-10. https://doi.org/10.1186/s12870-016-0771-y.

Viallard V, Poirier I, Cournoyer B, Haurat J, Wiebkin S, Ophel-Keller K, Balandreau J. *Burkholderia graminis* sp. nov., a rhizospheric *Burkholderia* species, and reassessment of *Pseudomonas phenazinium, Pseudomonas pyrrocinia* and *Pseudomonas glathei* as *Burkholderia*. *International Journal of Systematic and Evolutionary Microbiology* (1998) *48*(2):549-563. https://doi.org/10.1099/00207713-48-2-549.

Visser MB, Majumdar S, Hani E, Sokol PA. Importance of the ornibactin and pyochelin siderophore transport systems in *Burkholderia cenocepacia* lung infections. *Infection and Immunity* (2004) *72*(5):2850-2857. https://doi.org/10.1128/IAI.72.5.2850-2857.2004.

Wang C, Zhang SH, Wang PF, Hou J, Zhang WJ, Li W, Lin ZP. The effect of excess Zn on mineral nutrition and antioxidative response in rapeseed seedlings. *Chemosphere* (2009) *75*(11):1468-1476. https://doi.org/10.1016/j.chemosphere.2009.02.033.

Wang B, Mei C, Seiler JR. Early growth promotion and leaf level physiology changes in *Burkholderia phytofirmans* strain PsJN inoculated switchgrass. *Plant Physiology and Biochemistry* (2015) *86*:16-23. https://doi.org/10.1016/j.plaphy.2014.11.008.

Waqas M, Shahzad R, Khan AL, Asaf S, Kim YH, Kang SM, Bilal S, Hamayun M, Lee IJ. Salvaging effect of triacontanol on plant growth, thermotolerance, macro-nutrient content, amino acid concentration and modulation of defense hormonal levels under heat stress. *Plant Physiology and Biochemistry* (2016) *99*:118-125. https://doi.org/10.1016/j.plaphy.2015.12.012.

Weilharter A, Mitter B, Shin MV, Chain PS, Nowak J, Sessitsch A. Complete genome sequence of the plant growth-promoting endophyte *Burkholderia phytofirmans* strain PsJN (2011). https://doi.org/10.1128/JB.05055-11.

Yang HC, Im WT, Kim KK, An DS, Lee ST. *Burkholderia terrae* sp. nov., isolated from a forest soil. *International Journal of Systematic and Evolutionary Microbiology* (2006) *56*(2):453-457. https://doi.org/10.1099/ijs.0.63968-0.

You LX, Zhang RR, Dai JX, Lin ZT, Li YP, Herzberg M, Zhang JL, Al-Wathnani H, Zhang CK, Feng RW, Liu H. Potential of cadmium resistant *Burkholderia contaminans* strain ZCC in promoting growth of soy beans in the presence of cadmium. *Ecotoxicology and Environmental Safety* (2021) *211*:111914. https://doi.org/10.1016/j.ecoenv.2021.111914.

Zaman NR, Chowdhury UF, Reza RN, Chowdhury FT, Sarker M, Hossain MM, Akbor MA, Amin A, Islam MR, Khan H. Plant growth promoting endophyte *Burkholderia contaminans* NZ antagonizes phytopathogen *Macrophomina phaseolina* through melanin synthesis and pyrrolnitrin inhibition. *PloS one* (2021) 16(9):0257863. https://doi.org/10.1371/journal.pone.0257863.

Zhao S, Wei H, Lin CY, Zeng Y, Tucker MP, Himmel ME, Ding SY. *Burkholderia phytofirmans* inoculation-induced changes on the shoot cell anatomy and iron accumulation reveal novel components of *Arabidopsis*-endophyte interaction that can benefit downstream biomass deconstruction. *Frontiers in Plant Science* (2016)7: 24. https://doi.org/10.3389/fpls.2016.00024.

Zolg W, Ottow JCG. *Pseudomonas glathei* sp. nov., a new nitrogen scavenging rod isolated from acid lateritic relicts in Germany. *Zeitschrift für allgemeine Mikrobiologie* (1975) *15*(4):287-299. https://doi.org/10.1002/jobm.19750150410.

Chapter 4

Phosphorus Fertility Management in Field Crop Production

Mohammad Mirzaei Heydari[1] and Davey L. Jones[2]

[1]Department of Agronomy and Plant Breeding, Isfahan (Khorasgan) Branch, Islamic Azad University, Isfahan, Iran
[2]School of Natural Sciences, Bangor University, Bangor, Gwynedd Wales, UK

Abstract

Phosphorus (P) is a key element which supports root growth and is frequently a limiting factor in crop production. This has motivated a global interest in gaining a better understanding of its functions in the plant-soil-microbial system alongside the wider environment. The main aim of this research is to achieve improved P use efficiency and sustainable agricultural production via P management practices aimed at reducing P fertilizer inputs and alleviating its adverse impacts on the environment and public health. It is the objective of the present review to investigate the role of different P sources on P bioavailability and to explore its cycling in both plants and soil while also exploring its effects on plant growth and agricultural production. The prerequisite to such inquiries is a sound knowledge of P transformation and movement in soil and plant to be exploited toward enhancing P efficiency and improving fertilizer application management in agricultural production. Increasing soil available P at lower environmental costs and with less adverse impacts on plant communities may only be achieved through application of organic fertilizers and bio-fertilizers to reduce the addition of mineral-based P fertilizers. This review critically evaluates both the conditions and sources of P alongside its role in agroecosystem nutrient cycling and agricultural production.

In: Biofertilizers
Editor: Philip L. Bevis
ISBN: 979-8-89113-082-1
© 2023 Nova Science Publishers, Inc.

94   Mohammad Mirzaei Heydari and David L. Jones

Keywords: fertilization, nutrition, phosphorus, plant, soil

1. Introduction

Phosphorus (P) is an important element for all forms of life due to its participation in myriad of metabolic pathways and its role as an essential macronutrient in plant growth and crop production (Ziadi et al., 2013; Bünemann et al., 2016; Daly et al., 2015). The world population is projected to grow from 7.9 billion to ca. 10 billion by 2050 (Badgley et al., 2007; Bacci, 2012). To meet the corresponding increase in the demand for food, a matching increase should occur in crop production (Bindraban et al., 2002). One effective way to increase crop production per unit area is to improve fertilizer use efficiency (Zhu et al., 2018). This can only be achieved through advances in our fundamental understanding of soil fertility and plant nutrition. The expansion of fertilizer responsive varieties in the Green Revolution, together with the realization by farmers of the importance of fertilizers, has led to high levels of fertilizer use in agriculture (Cassman, 1999; Tilman et al., 2002; Colombi et al., 2017). Enhanced future demand for food may require even greater fertilizer inputs, but the increase in agricultural production required for our future food security must be achieved without endangering human health or the wider environment.

Knowledge of soil fertility is vital for the development of soil management systems that produce profitable crop yields while maintaining soil sustainability and environmental quality (Foth and Ellis, 1996; Gul and Whalen, 2016; Sharma et al., 2017). Soil fertility is a measure of the ability of soils to provide elements necessary for plant growth without reaching toxic concentrations of any element, while soils are classed as fertile when they are balanced providers of elements in a sufficiently labile or usable form to satisfy plant requirements (Foth and Ellis, 1996; Shahid et al., 2017).

The effectiveness of roots and associated rhizosphere organisms to acquire nutrients is extremely dependent on the soil's inherent fertility and the amount of exogenously supplied nutrients. It therefore follows that root growth and development can be significantly constrained when the soil is depleted in available nutrients (Zhang et al., 2010; Manschadi et al., 2014; Valença et al., 2017). It has been established that roots are able to enhance soil organic matter (and by association organic P reserves) by contributing nitrogen (via N_2 fixation), organic carbon (via rhizodeposition), and microbial biomass (e.g., via mycorrhizas)(Xie et al., 2014; Rasmussen et al., 2015;

Mundaa et al., 2016; Maji et al., 2017). Accordingly, an understanding of the factors that affect root growth and development is essential for improving nutrient cycling and P uptake by plants. Consequently, good plant growth and development and, thereby, desirable agricultural production, rely on the growth and development of the fine root system and associated symbionts (Mollier & Pellerin, 1999; Gao et al., 2010; Gahoonia et al., 1997; Fageria & Moreira, 2011; Shi et al., 2013).

P is one of the nutrients which is essential for crop growth and maximal utilization of other nutrients as nitrogen (N) (Brady and Weil, 2002; Mehrvarz et al., 2008). It is, however, one of the least available and mobile in soils (Hinsinger, 2001; Roy, 2017). Soil amendment with soluble chemical P fertilizers is a costly and potentially contaminating practice, especially if one considers the highly polluting mode of production of these fertilizers (Sharply et al., 2000; Meyer et al., 2017). Natural reserves of phosphate rock required for P fertilizer manufacture are also becoming rapidly depleted while there is no alternative to phosphate rock as a P source (Figure 1.1); hence, the importance of preserving the reserves for future generations (Cordell et al., 2009).

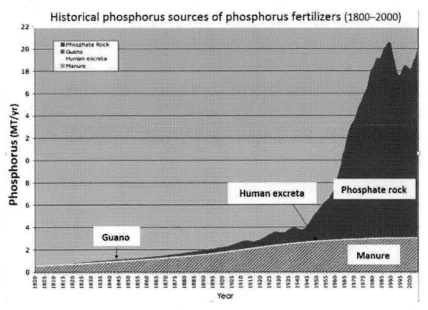

Figure 1.1. Consumption of phosphorus sources for use as fertilizers through time (1800–2000) (Reliability of data sources vary; hence, the data points for human excreta, guano, and manure should be interpreted as indicative rather than precise) (Cordell et al., 2009).

One potential option to decrease the hazardous impacts of P fertilizers on the environment while also maintaining high crop yields and farmer income is to replace the expensive soluble chemical P fertilizers with new, cheaper, more efficient bio-inoculants. These offer a potentially useful means of increasing nutrient uptake from the soil, thereby relaxing dependence on chemical inputs (Buckingham and Jasinski, 2004; Mirzaei-Heydari, 2013; Rodríguez-caballero et al., 2017).

2. Plant Nutrition and Soil Fertility

Soil fertility and plant nutrition are the two strongly related phenomena that directly affect not only the configuration and availability of nutrients in soil but also their transportation and uptake by roots as well as their utilization by plants (Foth and Ellis, 1996; Zornoza et al., 2017).

Nutrients should be available in sufficient and balanced quantities for optimum plant growth. Although soils serve as natural reserves of plant nutrients, they are mostly in forms unavailable to plants as only small portions are released during each growing season through biological activities and chemical processes (Chen, 2006; Bünemann et al., 2016). Moreover, available forms of nutrients are released too slowly to compensate for those taken up by crop plants and removed after harvest.

In order to increase soil nutrient supplies in forms readily available to plants, fertilizers need to be applied. Inorganic fertilizers (chemical fertilizers) and organic ones or bio-inoculants have their own advantages and disadvantages with respect to nutrient provision, crop growth, and their effects on the environment as well as human and animal health (Xie et al., 2011; Schoumans et al., 2014; Saikia et al., 2015). Optimized use of each fertilizer type to achieve a balanced nutrient management favourable to sound plant growth and healthy crop production presupposes proper consideration of the advantages and disadvantages.

2.1. Soil Fertility

Soil fertility depends on a wide range of physical, biological, and chemical factors. Soil is an active and complex system hosting living organisms interacting with particles of inorganic mineral and organic substances.

Microbial activity is strongly connected with nutrient availability and soil fertility (Garcia et al., 1994; Liu et al., 2016). A diverse range of roles is executed by soil that directly or indirectly sustain the world's human population. In food production, soil performs a vital function; it acts as a reservoir for water, a store of nutrients, a medium for root anchorage, and a filter for pollutants. Additionally, soils store approximately twice as much carbon (C) as the atmosphere and play a vital role in regulating the levels of atmospheric carbon dioxide (O'Donnell and Gorres, 1999; Weigel and Manderscheid, 2012; Saikia et al., 2015).

2.2. Plant Nutrition

Plants absorb nutrients, along with water, through their root hairs from a zone of approximately one centimetre around the root. For some nutrients, the mechanism of root uptake of soil nutrient solution involves transportation from more saturated to less saturated areas by mass flow (Chen et al., 2009; Zornoza et al., 2017). The efficacy of crop uptake of applied fertilizers is a determinant of fertilizer utilization efficiency. Enhanced nutrient absorption and utilization capacity of plants evidently entails enhanced efficiency of fertilizer application (Graham and Welch, 1996; Shenoy and Kalagudi, 2005). The anatomical and physiological characteristics of plants lead to great variations in optimal and critical nutrient values in different species or varieties, even under identical growing conditions (Drechsel and Zech, 1993). Crop roots and their associated symbionts play a key role in this nutrient uptake due to their ability to increase the radius of the zone from which nutrients can be absorbed (Ge et al., 2000). In contrast to N and K, phosphorus is relatively immobile in soil and reaches the root surface by diffusion, often moving only a few millimetres in one season.

2.3. Fertilizers

Fertilizers, widely used in agricultural systems on a global scale, may be classified into four types according to their production process:

1. Manufactured inorganic fertilizers (chemical fertilizers (CF));
2. Organic fertilizers (OF) including animal manures, biosolids, crop residues, compost, and green manures;

98   Mohammad Mirzaei Heydari and David L. Jones

3. Natural mineral fertilizers (rock phosphate); and
4. Bio-inoculants, including azotobacters, rhizobia, phosphate-solubilizing bacteria, mycorrhizas, and plant growth promoting rhizobacteria (Chand et al., 2006; Magadlela et al., 2017).

2.3.1. Chemical Fertilizers

Human and animal excreta were the initial empirically acknowledged fertilizers although only the Chinese applied some residues in an organized way until about 300 years ago (Igual, and Rodrigues-Barrueco, 2003; Jha et al., 2017). The experiments and research by the French scientists Antonie Lavoisier in 1774 and Boussingault in 1834 as well as those by the German chemist Liebig in 1840 laid the foundations of plant nutrition knowledge; these scientists undertook chemical analyses of plants and soils and succeeded in proving that chemical elements in plants came from soil and air. Von Liebig considered this theoretical base as the main advance within the fertilizer industry that led to the recognition of mineral elements as important ingredients in plant nutrition (Igual and Rodrigues-Barrueco, 2003; Bokhorst et al., 2017).

Chemical fertilizers have played important roles in the Green Revolution that not only ensured global food security but also transformed developing countries (Gyaneshwar et al., 2002; Adesemoye and Kloepper, 2009). However, intensive cropping and incessant use of high levels of chemical fertilizers frequently lead to an imbalance in soil nutrient supply, ultimately causing a decline in crop productivity (Nambiar, 1994; Maji et al., 2017; Ortas et al., 2019; Hammad et al., 2019).

2.3.2. Organic Fertilizers

Organic fertilizers (OF) can function as alternatives to mineral ones or bio-fertilizers to improve soil structure and microbial biomass. Hence, application of OF might raise crop yields while also minimizing CF application. This has recently motivated a growing demand by consumers for the so-called "organic" products (Garg and Bahl, 2008; Hammad et al., 2019), defined as good quality and safer foods grown without the use of manufactured chemical inputs. Organic farming relies on crop rotations, compost, animal manure, and plant residues to preserve soil fertility without the use of mineral fertilizers (Elfstrand et al., 2007). Moreover, application of organic matter has significant effects on soil such as improved physical, chemical, and biological properties as well as enhanced functional stability of the microbial community (Toyota and Kuninaga, 2006; Mundaa et al., 2016). Thus, organic manure can be used as a replacement for mineral fertilizers (Gupta et al., 1988; Wong et al., 1999; Naeem et al., 2006) to improve soil structure (Bin, 1983; Dauda et al., 2008) and the microbial biomass (Suresh et al., 2004).

It should not be, however, ignored that OFs when used alone might pose problems to both human health and the environment (Arisha and Bardisi, 1999). Additionally, OFs might be plagued with such disadvantages as low nutrient content per unit volume or mass (meaning that large amounts are required to provide sufficient nutrition for crops), rather slow rate of nutrient release for short-term production, and lack of certain plant nutrients needed for high crop yields (Mandal et al., 2007). This might be aggravated in most cases by the difficulty to procure the large quantities of OF required on farms.

2.3.3. Natural Mineral Fertilizers (Rock Phosphate)

Rock phosphate (RP) is a valuable material used as a P fertilizer (Ditta et al., 2018) and has in recent years received a growing attention by scholars and farmers alike (Bhatti and Yawar, 2010; Tarraf et al., 2017). As a natural, low-cost material that requires no particular processing, it has been accepted as a valuable P fertilizer, particularly for acid soils (Goenadi et al., 2000; Stamford et al., 2007). Its solubility in non-acidic (calcareous) soils is, however, rather low while the major problem for its direct application to farm soils is the low rate of P release into the soil solution (Khasawneh and Doll, 1978; McLaughlin et al., 1990; Kpomblekou and Tabatabai, 2003; Rajan and Watkinson, 1992). Nevertheless, plants are reportedly capable of solubilizing P from RP through their root acid exudates (Bolan et al., 1990; Hinsinger, 1998; Liu et al., 2007) while crop varieties have also been reported to exhibit different acidification levels of their rhizospheres (Flach et al., 1987; Wang et al., 2015). Meanwhile, PR effects on plant growth might be manipulated by such practices as decreasing particle size, acidulating RP by artificial or natural organic acids (Amarasinghe et al., 2022), mixing with chemical materials to produce acids (e.g., elemental sulphur, pyrites, $(NH_4)_2SO_4$, or (NH_4NO_3) in non-acidic soils, sulphur oxidation, and use of PSM (Ditta et al., 2018). Calcareous and alkaline soils are capable of increasing phosphorus availability from RP by attenuating soil reactions (Ditta et al., 2018; Singh and Amberger, 1998; Stamford et al., 2003; Sagoe et al., 1998; Hammond et al., 1986; Lewis et al., 1997; Rajan and Watkinson, 1992).

2.3.4. Bio-Inoculants

Bio-inoculants may be considered as a subtype of bio-fertilizers that contain living organisms capable of enhancing nutrient acquisition by the host plant through their continuous presence within the plant's rhizosphere (Chen, 2006; Trabelsi et al., 2017). Many plants have benefited from symbiotic associations with micro-organisms under P-deficient and drought conditions (Zaheer et al., 2019). These associations might result either in better uptake of the available P in soil or in rendering unavailable P sources accessible to the plant. Bio-inoculants utilize single or multiple strains of naturally occurring microorganisms to change essential elements, such as P, from unavailable to available forms via biological and chemical processes (Richardson, 1994: Zheng et al., 2011; Shahzad et al., 2017; Maji et al., 2017). Various claims have often been made about their beneficence and ability to promote plant growth, thereby relaxing part of the need for chemical fertilizers (Rai, 2006; Trabelsi et al., 2017; Ditta et al., 2018).

Bio-inoculants are categorized into the following five main groups:

1. Arbuscular mycorrhizas (AM);
2. Phosphorus solubilizing bacteria (PSB);
3. Plant growth promoting rhizobacteria (PGPR);
4. Azotobacters and azospirillum; and
5. *Rhizobium* spp.

2.3.4.1. Advantages of Bio-Inoculants

There is no shortage of literature on the beneficial effects of bio-inoculants. Rai (2006), Mirzaei Heydari et al. (2009), and Lin et al. (2020) noted that they not only promote plant growth and reduce chemical inputs but easily transport water and nutrients from the rhizosphere for better absorption by plants. In their field study of PSB and AM application, Singh et al. (2011) and Mirzaei Heydari et al. (2010) found that, compared to mineral phosphorus, bio-inoculants were cheaper, increased phosphorus uptake by the plant, helped increase yield and yield components, and reduced chemical fertilizer application. Buckingham and Jasinki (2004) described bio-fertilizers as being relatively easy to produce (El Maaloum et al., 2020). Sinha (1998) and Mirzaei Heydari et al. (2011) maintained that bio-fertilizer application was an efficient means of improving soil fertility, enhancing nutrient absorption, and helping plants absorb trace elements from around the root zones. They also noted that they are ecologically safer than mineral fertilizers while they rely on less costly equipment and production processes compared to chemical fertilizers.

Weller (1988) and Singh and Ram (2005) investigated the effects of bio-fertilizer application combined with reduced quantities of artificial compounds to find that the practice not only decreased the adverse environmental impacts associated with chemical fertilizers but also enhanced, or at least maintained, yields at a sustainable level by assisting plant growth, promoting bacteria, and optimising fertilizer application rates. It may, thus, be concluded that an integrated plant nutrient-management system that considers decreased inputs of chemical fertilizers combined with the application of organic inputs and/or bio-fertilizers represents an alternative and more sustainable strategy for crop nutrition (Oberson et al., 1996; Gunapala and Scow, 1998; Imtiaz et al., 2016; Munda et al., 2016; Rodríguez-caballero et al., 2017; Mitra et al., 2020).

3. Phosphorus (P)

Henning Brandt of Germany discovered phosphorus in 1557. The name is derived from the Greek words *phos* (light) and *phorus* (bringing) denoting 'light production' (Boyd, 1990) due to its luminescent properties and spontaneous combustion in air. Additionally, outside of agriculture, P is extensively used in the production of detergents, explosives, matches, and flares (Massey, 2010).

P does not occur in plenty in natural ecosystems although it is a major plant nutrient and plays a fundamental role in agriculture and biogeochemical cycles. Found in all parts and organs of living organisms, phosphorus forms numerous covalent organo-phosphorus compounds and binds to C, N, O, Al, Fe, and Ca. It is involved in the primary transfer of radiant electromagnetic energy to chemical photosynthetic energy and helps sustained plant growth (Bahl et al., 1997; Garg and Bahl, 2008; Feng et al., 2016; Imtiaz et al., 2016).

3.1. Phosphorus Cycle

Phosphorus is one of the lowest biologically available nutrients and yet is abundant in nature. It is similar to other mineral nutrient cycles in that it exists in minerals, soils, water, and living organisms (Schachtman et al., 1998; Ashley et al., 2011). Elemental P is highly reactive and violently reacts with oxygen when exposed to air (Kirkby & Johnston 2008). In natural systems (such as water and soil), it exists as phosphate ions formed where each P atom binds to four oxygen atoms (Takahashi & Anwar 2007). Orthophosphate is the simplest form of phosphate that exists in water as PO_4^{3-}, in acidic environments as $H_2PO_4^{1-}$, and as HPO_4^{2-} in alkaline media (Fardeau & Guiraud 1996; Hamadoun et al., 2016).

P is taken up from soil by plants, which are in turn consumed by humans and animals, and returned to soil as organic residues, animal debris, and decay material in soil (Figure 3.1). The organic P for plant material in soil is released in the form of inorganic phosphate (mineralization) by soil microorganisms (Rodríguez & Fraga 1999; Mkhabela & Warman, 2005). P can be lost through runoff, soil erosion, and leachates into groundwater (Dawson & Hilton, 2011). There exist phosphate salts that are not highly soluble in water; hence, the majority of P exists in the solid form in natural systems (Vazquez et al., 2000).

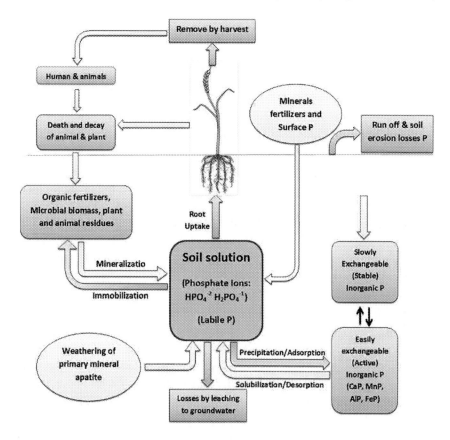

Figure 3.1. Phosphorus transformation and movement in soil and plant.

3.2. Soil Phosphorus

Phosphorus is the major essential nutrient element (after nitrogen) that most often limits agricultural production. Compared to other main nutrients, P is the least mobile and available to plants under most soil conditions (Schachtman et al., 1998; Hinsinger, 2001). However, P is the 11th most abundant element in the soil, with typical concentrations in the range of 0.1- 3 mg P kg soil (Hedley et al., 1995; Mengel, 1997).

Although farm soils often contain high total quantities of P, the element frequently occurs in unavailable forms while a major portion of the available form may be far from plant roots (i.e., outside of the rhizosphere). More than 80% of soil P content becomes immobilized through transformation to organic forms, adsorption, and precipitation to become unavailable for plant uptake (Holford, 1997). While dissolved inorganic P is essential for plant uptake, a portion of the dissolved organic P may also be utilized by plants (Turner and Haygarth, 2000b; Alvarado et al., 2019).

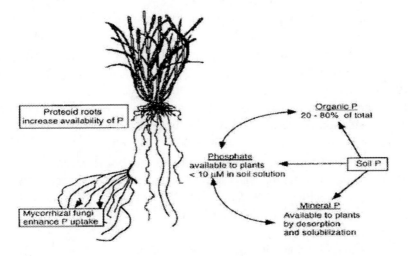

Figure 3.2. Plant acquisition of soil P (Adapted from Schachtman et al., 2001).

P in soil is present in different forms of organic (P_o) and inorganic (P_i) minerals (at least 170 different minerals contain phosphates), both of which are important sources for microbes and plant uptake (Rao et al., 1999; Richardson, 2003). Inorganic phosphate includes free P ions in the soil solution, unstable P bound to soil particles (mainly clays, salts of insoluble inorganic P_i including aluminium phosphate and iron phosphate in acidic soils or calcium phosphate in alkaline soils, for instance), and components of complex organic compounds in the organic material of soil (Rao et al., 1999; Shenoy & Kalagudi 2005). More than 90% of the whole P in the soil-plant-animal cycle resides in soil while approximately 10% is found in the residual biological cycle (Ozanne, 1980; Oberson and Joiner, 2005; Turner, 2007; Kirkby and Johnston, 2008).

3.3. Phosphorus in Plants

Phosphorus is a necessary nutrient for plant growth, development, and fertility. It is a component of key molecules like phospholipids in cell membranes, adenosine triphosphate (ATP), adenosine diphosphate (ADP) (both being essential in energy storage and transfer reactions), deoxyribonucleic acids (DNA), and ribonucleic acids (RNA) (the two nucleic acid components of genetic information). It also helps in the early ripening of plants, decreasing grain moisture, and enhancing crop quality (Sharma, 2002).

Clearly, plants cannot grow without a reliable supply of this vital nutrient (Thedorou and Plaxton, 1993). In natural and agricultural ecosystems, however, availability of P regularly limits plant growth (Agren 1988; Vance et al., 2003; Gusewell, 2004). This is because plants can only acquire P via root hairs and their connected arbuscular mycorrhizal fungi (AMF) taking up P (orthophosphate anions), mostly HPO_4^{2-} and to a lesser extent $H_2PO_4^{1-}$, from the soil solution (Bieleski, 1973; Ullrich-Eberius et al., 1984; Furihata et al., 1992; Schachtman et al., 1998; White, 2003). Orthophosphate ions exist in the soil solution at extremely low concentrations (<10 μM) because of the low solubility of Pi salt products (Hedley et al., 1995; Marschner 1995). In addition, the quantity of P in the soil solution of the rhizosphere declines very rapidly during crop growth despite the slow replenishment of P from soil P sources (Barber, 1995).

Compensation for soil P deficit and enhancement of P uptake efficiency can help plant P absorption. Enhanced P availability may be achieved through such means as changing the rhizosphere so that bio-molecules capable of releasing P from organic-P complexes or metallic-P compounds are set free (Marschner, 1995; Johnson et al., 1996), modifying soil structure to increase root-soil contact area to increase root absorption area (Lynch,1995), exploiting soil microbes like arbuscular mycorrhizal fungi (Schachtman et al., 1998; Trabelsi et al., 2017), and enhancing phosphate production (Bariola et al.,

1994). In addition, the application of phosphorus fertilizers can be used to raise soluble P concentrations in the rhizosphere (Figure 3.3).

P concentrations in plant tissues range between 0.2% and 0.5% of total dry matter where P is not a growth limiting factor (Sanchez, 2007). Plants absorb P only as P_i ions from the soil solution (Holford, 1997). It is, therefore, essential for plant nutrition that P sources are mineralized or solubilised in order to release soluble P (Mengel, 1997; Ashley et al., 2011).

As already mentioned, the properties of both soil and plant interactively affect P acquisition. Availability of P for plants depends on such soil processes as P solubility or sorption, mineralization or immobilization, root-soil contact surface area, and P transport. Also, the release and availability of P in soil is dependent on soil pH, with the highest rates of P uptake reportedly occurring between at pH levels of 5.0 and 6.0 (Ullrich-Eberius et al., 1984; Furtihata et al., 1992; Zhu et al., 2018).

3.4. Chemical Phosphorus Fertilizers

Lawes and Gilbert produced superphosphate in 1842 for the first time from compressed bones and mined phosphate by treating with sulphuric acid (Igual and Rodrigues-Barrueco, 2007). This has become the major form of mineral fertilizer supplying soluble P to agricultural systems.

Application of chemical phosphorus fertilizers (CPF) has become part and parcel of agricultural production due to the need for increased crop production and the limited availability of P to plant roots in two thirds of the agricultural soils of the world. The main active component of P fertilizers is phosphorus obtained from RP. Phosphorus fertilizers commonly used in agricultural systems include triple superphosphate (TSP) (17-23% P; 44-52% P_2O_5), single superphosphate (SSP) (7-9.5% P; 16-22% P_2O_5), diammonium phosphate (DAP) (20% P; 46% P_2O_5), and monoammonium phosphate (MAP) (48-61% P_2O_5). RP is the raw input to the chemical processes to produce P fertilizers and process of manufacturing P fertilizers involves milling of rock phosphate and treatment with acid to produce phosphoric acid followed by heating to drive off water (Batjes, 1997; Tunney et al., 2003; Chen et al., 2006; Roselli et al., 2009).

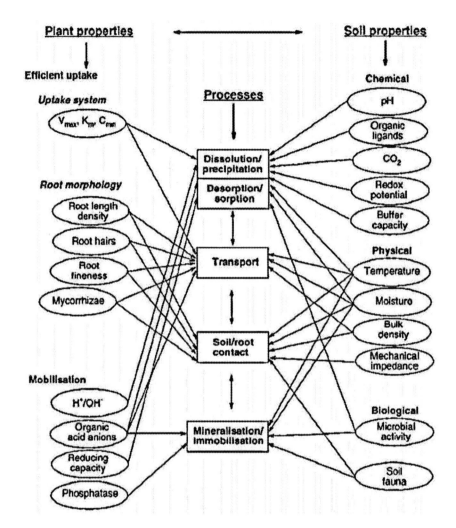

Figure 3.3. Processes governing acquisition of soil and fertilizer P by crops as affected by plant and soil properties (Adapted from Horst et al., 2001).

4. Phosphorus and Future Prospects

With the growing demand for agricultural production and food, the highest production on a global scale is expected to happen in the decades to come (Cordell et al., 2009; Shen et al., 2011). Given the importance of P to life, soil fertility, plant nutrition, and food security (Chen and Graedlel, 2016; Roy,

2017), a sustainable management of P in agriculture systems is needed to ensure sustained supply of this vital element (Bindraban et al., 2020).

One of the future challenges soil nutrition and plant fertilization will expectedly face in future include depletion of P sources and available P deficiency in agricultural soils. The development of new and effective management techniques to ensure sustainable agriculture heavily relies on an understanding of the relationship between soil fertility and plant nutrition to address the problems associated with P deficiency and P overabundance in soil.

Conclusion

P is the second most important micronutrient required for crop growth and agricultural production (Xiaozhu et al., 2017). Application of P fertilizers is the most common way to address P deficiency and to enhance soil fertility to ensure satisfactory crop production. However, P resources are non-renewable and the majority of agricultural soils suffer from limited P supplies. The situation may be worsened in the absence of a proper P fertilization management to give rise to dire consequences for the environment, food security, and public health. The shortages might be compensated for and the problems remedied if a sound knowledge and adequate understanding are gained of P transformation and movement in soil and plant to be exploited in designing application management and enhancing P utilization efficiency in agricultural production. Increasing available P in soil for uptake by plant communities at lower environmental costs presupposes application of organic fertilizers and bio-fertilizers, which undoubtedly guarantee a matching decline in chemical P fertilizer application.

References

Adesemoye, A. O., Kloepper, J. W., 2009. Plant-microbes interactions in enhanced fertilizer-use efficiency. *Applied microbiology and biotechnology*. 85, 1–12.

Alvarado, J. S., McCray, J. M., Erickson, J. E., Sandhu, H. S. and Bhadha, J. H., 2019. Sugarcane biomass yield response to phosphorus fertilizer on four mineral soils as related to extractable soil phosphorus. *Communications in Soil Science and Plant Analysis,* 50(22), pp. 2960-2970.

Amarasinghe, T., Madhusha, C., Munaweera, I. and Kottegoda, N., 2022. Review on Mechanisms of Phosphate Solubilization in Rock Phosphate Fertilizer. *Communications in Soil Science and Plant Analysis*, pp. 1-17.

Arisha, H. M., Bradisi, A., 1999. Effect of mineral fertilizers and organic fertilizerss on growth, yield and quality of potato under sandy soil conditions. *Zagazig J. Agric. Res.,* 26, 391–405.

Bacci, M. L., 2012. *A Concise History of World Population* 5, Revised., Chicester/GB: John Wiley and Sons Ltd.

Badgley, C., Moghtader, J., Quintero, E., Zakern, E., Chapell, J., Avilés-Vázquez, K., Samulon, A., Perfecto, I., 2007. Organic agriculture and the global food supply. *Renewable Agriculture and Food Systems*. 22, 86–108.

Bhatti, T. M., and W. Yawar. 2010. Bacterial solubilization of phosphorus from phosphate rock containing sulfur-mud. *Hydrometallurgy.* 103, 54–59.

Bin, J., 1983. Utilization of green manure for raising soil fertility in China. *Soil Sci.,* 135: 65– 9 *Agric. Res.,* 25: 1087–101.

Bindraban, P. S., Dimkpa, C. O. and Pandey, R., 2020. Exploring phosphorus fertilizers and fertilization strategies for improved human and environmental health. *Biology and Fertility of Soils,* 56(3), pp. 299-317.

Bokhorst, S., Kardol, P., Bellingham, P. J., Kooyman, R. M., Richardson, S. J., Schmidt, S., and Wardle, D. A., 2017. Responses of communities of soil organisms and plants to soil aging at two contrasting long-term chronosequences. *Soil Biology and Biochemistry*, 106, 69–79.

Bolan, N. S., 1991. A critical review on the role of mycorrhizal fungi in the uptake of phosphorus by plants. *Plant and Soil*, 134, 189–207.

Brady, N. C., and Weil, R. R., 2002. *The nature and properties of soils*, 13th edition. Upper Saddle River, NJ: Prentice Hall.

Buckingham, D., and S. Jasinski, 2004. *Phosphate Rock Statistics 1900–2002*. US Geological Survey.

Bünemann, E. K., Augstburger, S., Frossard, E., 2016. Dominance of either physicochemical or biological phosphorus cycling processes in temperate forest soils of contrasting phosphate availability. *Soil Biology & Biochemistry*, 101, 85–95.

Byngaard, N., M. L. Cabrera, K. A. Jarosch, 2016. Phosphorus in the coarse soil fraction is related to soil organic phosphorus mineralization measured by isotopic dilution. *Soil Biology and Biochemistry*, 96, 107–118.

Cassman KG, Peng S, Olk DC, Ladha JK, Reichardt W, Dobermann A, Singh U (1998) Opportunities for increasing nitrogen use efficiency from improved resource management in irrigated rice systems. *Field Crop Res.* 56, 7–38.

Chen, J. H., 2006. The combined use of chemical and organic fertilizers and/or biofertilizer for crop growth and soil fertility. International Workshop on Sustained Management of the Soil-Rhizosphere System for Efficient Crop Production and Fertilizer Use, 16 – 20 October 2006, Department of Soil and Environmental Sciences, National Chung Hsing University, Taiwan, pp. 1–11.

Chen, M., Graedel, T. E., 2016. A half-century of global phosphorus flows, stocks, production, consumption, recycling, and environmental impacts. *Global Environ. Change,* 36, 139-152.

Chen, J. Y., Xu, L., Cai, Y. L. and Xu, J. 2009. Identification of QTLs for phosphorus utilization efficiency in maize (*Zea mays* L.) across P levels. *Euphytica*. 167, 245–252.

Chand, S., Anwar, M., Patra, D. D., 2006. Influence of long-term application of organic and inorganic fertilizer to build up soil fertility and nutrient uptake in mint mustard cropping sequence. *Communications in Soil Science and Plant Analysis*. 37, 63-76.

Colombi, T., Braun, S., Keller, T. and Walter, A., 2017. Artificial macropores attract crop roots and enhance plant productivity on compacted soils. *Science of the Total Environment, The*, 574, 1283–1293.

Cordell, D., Drangert, J. and White, S. 2009. "The story of phosphorus: Global food security and food for thought," *Global Environmental Change*, 19, 292-305.

Daly, K., Styles, D., Lalor, S., Wall, D. P., 2015. Phosphorus sorption, supply potential and availability in soils with contrasting parent material and soil chemical properties. *Eur. J. Soil Sci.* 66, 792–801.

Dauda, S. N., F. A. Ajayi and E. Ndor, 2008. Growth and yield of water melon (*Citrullus lanatus*) as affected by poultry manure application. *J. Agric. Soc. Sci.*, 4: 121–421.

Ditta, A., Imtiaz, M., Mehmood, S., Rizwan, M. S., Mubeen, F., Aziz, O., Qian, Z., Ijaz, R. and Tu, S., 2018. Rock phosphate-enriched organic fertilizer with phosphate-solubilizing microorganisms improves nodulation, growth, and yield of legumes. *Communications in Soil Science and Plant Analysis*, 49(21), pp. 2715-2725.

Drechsel, P., Zech, W. 1993. Mineral nutrition of tropical trees. In *Tropical Forestry Handbook*. Vol. 1, (Ed. L. Pancel). Spring-Verlag, 516-567.

Elfstrand S, Hedlund K, and Martensson A. 2007. Soil microbilcommunity composition and function after 47 years of continous green manuring. *Applied Soil Ecology*, 35, 610-621.

El Maaloum, S., Elabed, A., Alaoui-Talibi, Z. E., Meddich, A., Filali-Maltouf, A., Douira, A., Ibnsouda-Koraichi, S., Amir, S. and El Modafar, C., 2020. Effect of arbuscular mycorrhizal fungi and phosphate-solubilizing bacteria consortia associated with phospho-compost on phosphorus solubilization and growth of tomato seedlings (Solanum lycopersicum L.). *Communications in Soil Science and Plant Analysis*, 51(5), pp.622-634.

Fageria, N. K. and Moreira, A., 2011. Chapter Four - The Role of Mineral Nutrition on Root Growth of Crop Plants. In D. L. S. B. T. A. in *Agronomy,* ed. Academic Press, pp. 251–331.

Flach, E. N., Quak, W., van Diest, A., 1987. A comparison of the rock phosphate mobilizing capacities of various crop species. *Trop. Agric.* 64, 347–352.

Foth, H. D. and Ellis, B. G., 1996. *Soil fertility*. (Book), 2nd Ed. Lewis Pub. New York.

Gahoonia, T. S., Care, D. and Nielsen, N. E., 1997. *Root hairs and phosphorus acquisition of wheat and barley cultivars*, pp. 181–188.

Gao, Y., Duan, A., Qiu, X., Liu, Z., Sun, J., Zhang, J. and Wang, H., 2010. Distribution of roots and root length density in a maize/soybean strip intercropping system. *Agricultural Water Management*, 98, 199–212.

García, T. Hernández, F. Costa and B. Ceccanti, 1994. Bichemical parameters in soils regenerated by the addition of organic wastes. *Waste Manag. Res*. 12. 457–466.

Garg, S. & Bahl, G. S. 2008. Phosphorus availability to maize as influenced by organic manures and fertilizer P associated phosphatase activity in soils. *Bioresource technology.* 99, 5773-5777.

Ge, Z., Rubio, G., Lynch, J. P. 2000. The importance of root gravi tropism for inter-root competition and phosphorusaquisition efficiency: Resultsfrom a geometric simulation model. *Plant and Soil*, 218, 159-171.

Goenadi, D. H., Siswanto and Y. Sugiarto. 2000. Bioactivation of poorly soluble phosphate rocks with a phosphorus-solubilizing fungus. *Soil Sci. Soc. Am. J.*, 64, 927-932.

Graham, R. D., Welch, R. M., 1996. Breeding for staple food crops with high micronutrient density, agricultural strategies for micronutrients: working paper 3. Washington DC7 International Food Policy Research Institute. 1– 72.

Gupta, A. P., Antil, S. R., Narwal, P. R., 1988. Effect of farmyard manure on organic carbon, available N and P contents of soil during different periods of wheat growth. *J. Indian Soil Sci.,* 36, 269–273.

Gul, S., Whalen, J. K., 2016. Soil Biology & Biochemistry Biochemical cycling of nitrogen and phosphorus in biochar-amended soils. *Soil Biology and Biochemistry,* 103, 1-15.

Gyaneshwar, P., Naresh Kumar, G., Parekh, L. J., Poole, P. S., 2002. Role of soil microorganisms in improving P nutrition of plants. *Plant and Soil*. 245, 83–93.

Hammad, H. M., Khaliq, A., Abbas, F., Farhad, W., Fahad, S., Aslam, M., Shah, G. M., Nasim, W., Mubeen, M. and Bakhat, H. F., 2020. Comparative effects of organic and inorganic fertilizers on soil organic carbon and wheat productivity under arid region. *Communications in Soil Science and Plant Analysis*, 51(10), pp. 1406-1422.

Hammond, L. L., Chien, S. H., Mokwunye, A. U., 1986. Agronomic value of unacidulated and partially acidulated phosphate rocks indigenous to the tropics. *Adv. Agron.* 40, 89–140.

Hinsinger, P., 2001. Bioavailability of soil inorganic P in the rhizosphere as affected by root- induced chemical changes: a review. *Plant Soil* 237, 173–195.

Hinsinger, P. 1998. How do plant roots acquire mineral nutrients? Chemical processes involved in the rhizosphere. *Adv. Agron.* 64, 225–265.

Igual, J., Rodriguez-Barrueco, C., 2007. Fertilizers, food and environment. first International Meeting on Microbial Phosphate Solubilization Developments in plant and soil. *Soil Sciences.* 102, 199-202.

Khasawneh, F. E. and Doll, E. C. 1978. The use of phosphate rock for direct application to soils. *Adv. Agron.* 30, 159–206.

Kpomblekou-A, K. & Tabatabai, M. a., 2003. Effect of low-molecular weight organic acids on phosphorus release and phytoavailabilty of phosphorus in phosphate rocks added to soils. *Agriculture, Ecosystems & Environment*, 100, 275–284.

Lewis, D. R., McGechan, M. B., 1998. Simulating field-scale nitrogen management scenarios involving fertilizer and slurry applications. Paper 98-E-057, AgEng98, International Conference, Oslo.

Li H. G., Shen J. B., Zhang F. S., and Lambers H. 2010. Localized application of soil organic `matter shifts distribution of cluster roots of white lupin in the soil profile due to localized release of phosphorus. *Annals of Botany* 105, 585–593.

Phosphorus Fertility Management in Field Crop Production

Lin, Y., Watts, D. B., Kloepper, J. W., Feng, Y. and Torbert, H. A., 2020. Influence of plant growth-promoting rhizobacteria on corn growth under drought stress. *Communications in Soil Science and Plant Analysis*, 51(2), pp. 250-264.

Liu, J., Liao, W., Zhang, Z., Zhang, H., Wang, X., Meng, N., 2007, Effect of Phopshate Fertilizer and Manure on Crop Yield, Soil P Accumulation, and the Environmental Risk Assessment, *Agricultural Sciences in China*, 6, 1107-1114.

Liu, T., Chen, X., Hu, F., Ran, W., Shen, O., Li, H., Whalen, J. K., 2016. Agriculture, Ecosystems and Environment Carbon-rich organic fertilizers to increase soil biodiversity : Evidence from a meta-analysis of nematode communities. *Agriculture, Ecosystems and Environment*, 232, 199–207.

Jha, S. K., Gao, Y., Liu, H., Huang, Z., Wang, G., Liang, Y. and Duan, A., 2017. Root development and water uptake in winter wheat under different irrigation methods and scheduling for North China. *Agricultural Water Management*, 182, 139–150.

Magadlela, A., Beukes C., Venter, F., Steenkamp, E. and Valentine. A., 2017. Does P de fi ciency affect nodule bacterial composition and N source utilization in a legume from nutrient-poor Mediterranean-type ecosystems ? *Soil Biology and Biochemistry*. 104, 164–174.

Maji, D., Misra, P., Singh, S., Kalra, A., 2017. Humic acid rich vermicompost promotes plant growth by improving microbial community structure of soil as well as root nodulation and mycorrhizal colonization in the roots of Pisum sativum. *Applied Soil Ecology*, 110, 97–108.

Mandal A., Patra A. K., Singh D., Swarup A., Ebhin Masto, R., 2007. Effect of long-term application of manure and fertilizer on biological and biochemical activities in soil during crop development stages. *Biores. Technol.*, 98, 3585–3592.

Manschadi, A. M., H. P. Kaula, J. Vollmannb, J. Eitzingerc and W. Wenzel. 2014. Reprint of "Developing phosphorus efficient crop varieties an interdisciplinary research framework." *Field Crops Research*, 165, 49–60.

McLaughlin, M. J., Baker, T. G., James, T. R., Rundle, J. A., 1990. Distribution and forms of phosphorus and aluminium in acidic topsoils under pastures in south eastern Australia. *Australian Journal of Soil Research*. 28, Eds T Lelly, E Balzs and M Tepler. Austria: Facultas.

Mehrvarz, S., Chaichi, M. R. and Alikhani, H. A., 2008. Effects of phosphate solubilizing microorganisms and phosphorus chemical fertilizer on yield and yield components of Barely (Hordeum vulgare L.). *Am-Euras. J. Agric. & Environ. Sci*. 3, 822-828.

Mitra, D., Anđelković, S., Panneerselvam, P., Senapati, A., Vasić, T., Ganeshamurthy, A. N., Chauhan, M., Uniyal, N., Mahakur, B. and Radha, T. K., 2020. Phosphate-solubilizing microbes and biocontrol agent for plant nutrition and protection: current perspective. *Communications in Soil Science and Plant Analysis*, 51(5), pp. 645-657.

Mirzaei Heydari, M., 2013. The role of bio-inoculants on phosphorus relations of barley. PhD Thesis, Bangor University, Wales, United Kingdom, 193 pp.

Mirzaei Heydari, M., Brook, R. M. & Jones, D. L., 2019. The Role of Phosphorus Sources on Root Diameter, Root Length and Root Dry Matter of Barley (Hordeum vulgare L.). *Journal of Plant Nutrition*. Vol. 42, No. 1, pp. 1–15.

112 Mohammad Mirzaei Heydari and David L. Jones

Mirzaei Heydari, M., Brook, R. M., Withers, P., Jones, D. L., 2011. Mycorrhizal infection of barley roots and its effect upon phosphorus uptake, *Aspects of Applied Biology*. 109, 137-142.

Mirzaei Heydari, M., R. Brook, P. Withers, and M. D. Keerio, 2010. The role of biofertilizers in sustainable agricultural systems: phosphorus. *International Conference on Bioscience, Biotechnology, and Biochemistry*, Italy, Rome, 28-30 April 2010.

Mirzaei Heydari, M., A. Maleki, R. Brook and D. L. Jones, 2009. Efficiency of Phosphorus Solubilising Bacteria and Phosphorus Chemical Fertilizer on Yield and Yield Components of Wheat cultivar (Chamran). *Aspects of Applied Biology* 98, 1-6.

Mollier, A. and Pellerin, S., 1999. *Maize root system growth and development as influenced by phosphorus deficiency*, 50, 487–497.

Mundaa, S., Shivakumar, B. G. Rana, D. S., Gangaiah, B., Manjaiah, K. M., Dass, A., Layek, J., Lakshman, K., 20016. Inorganic phosphorus along with biofertilizers improves profitability and sustainability in soybean (Glycine max)– potato (Solanum tuberosum) cropping system. *Journal of the Saudi Society of Agricultural Sciences*, pp. 4–10.

Meyer, G., Bünemann, E. K., Frossard, E., Maurhofer, M., Mader, P., and Oberson, A., 2017. G*ross phosphorus fluxes in a calcareous soil inoculated with Pseudomonas protegens CHA0 revealed by 33 P isotopic dilut*ion. 104, 81–94.

Mundaa, S., Shivakumar, B. G. Rana, D. S., Gangaiah, B., Manjaiah, K. M., Dass, A., Layek, J., Lakshman, K., 20016. Inorganic phosphorus along with biofertilizers improves profitability and sustainability in soybean (Glycine max)– potato (Solanum tuberosum) cropping system. *Journal of the Saudi Society of Agricultural Sciences*, pp. 4–10.

Naeem, M., J. Iqbal and M. A. A. Bakhsh, 2006. Comparative Study of Inorganic Fertilizers and Organic Manures on Yield and Yield Components of Mungbean (Vigna radiate L.). *J. Agric. Soc. Sci.*, 2, 227–229.

Ortas, I. and Islam, K. R., 2018. Phosphorus fertilization impacts on corn yield and soil fertility. *Communications in Soil Science and Plant Analysis*, 49(14), pp. 1684-1694.

Rai M. K. 2006. *Handbook of Microbial Biofertilizers*. Binghamton, New York: Haworth Press.

Rajan, S. S. S., Watkinson, J. H., 1992. Unacidulated and partially acidulated phosphate rock: agronomic effectiveness and the rates of dissolution of phosphate rock. *Nutr. Cycle Agroecosys.* 33, 267–277.

Rasmussen, I. S., Dresbøll, D. B. and Thorup-Kristensen, K., 2015. Winter wheat cultivars and nitrogen (N) fertilization—Effects on root growth, N uptake efficiency and N use efficiency. *European Journal of Agronomy*, 68, 38–49.

Richardson, A. E., 1994. Soil micro-organisms and phosphate availability. In *Soil Biota Management in Sustainable Agriculture*. Eds. CE Pankhurst, BM Doube, VVSR Gupts and PR Grace pp. 50–62. CSIRO, Melbourne, Australia.

Rodríguez-caballero, G., F. Caravaca, A. J. Fernández-González, M. M. Alguacil, M. Fernández-López, A. Roldán. 2017. Arbuscular mycorrhizal fungi inoculation mediated changes in rhizosphere bacterial community structure while promoting revegetation in a semiarid ecosystem. 585, 838–848.

Phosphorus Fertility Management in Field Crop Production 113

Roy, E. D., 2017. Phosphorus recovery and recycling with ecological engineering : A review. *Ecological Engineering*, 98, 213–227.

Saikia, P., Bhattacharya, S. S., Baruah, K. K., 2015. Organic substitution in fertilizer schedule: Impacts on soil health, photosynthetic efficiency, yield and assimilation in wheat grown in alluvial soil. *Agriculture, Ecosystems & Environment*. 203, 102–109.

Sagoe, C. I., Ando, T., Kouno, K., Nagaoka, T., 1998. Effects of organic acid treatment of phosphate rocks on the phosphorus availability to Italian ryegrass. *Soil Sci. Plant Nutr.* 43, 1067–1072.

Schoumans, O. F., Chardon, W. J., Bechmann, M. E., Gascuel-Odoux, C., Hofman, G., Kronvang, B. Rubæk, G. H., Ulén, f,B., Dorioz, J. M., 2014. Mitigation options to reduce phosphorus losses from the agricultural sector and improve surface water quality: A review. *Science of the Total Environment*, 468–469, 1255–1266.

Shahid, M., Shamshad, S., Rafiq, M., Khalid, S., Bibi, I., Niazi, N. K., Dumat, C., Rashid, M. I., 2017. Chromium speciation, bioavailability, uptake, toxicity and detoxi fi cation in soil-plant system : A review. *Chemosphere*. 178, 513-533.

Shahzad, S. M., Arif, M. S., Riaz, M., Ashraf, M., Yasmeen, T., Zaheer, A., Bragazza, L. and Bulttler A. 2017. Soil & Tillage Research. Interaction of compost additives with phosphate solubilizing rhizobacteria improved maize production and soil biochemical properties under dryland agriculture. *Soil & Tillage Research*, 174, 70–80.

Sharma, B., Sarkar, A., Singh, P., Singh R. P., 2017. Agricultural utilization of biosolids : A review on potential effects on soil and plant grown. *Waste Management*, 64, 117–132.

Sharpley, A., Foy, B. and P. Withers, 2000. Practical and innovative measures for the control of agricultural phosphorus losses to water: an overview. *Journal of Environmental Quality*. 29, 1-9.

Shenoy, V. V., Kalagudi, G. M., 2005. Enhancing plant phosphorus use efficiency for sustainable cropping. *Biotechnology advances*, 23, 501–513.

Shen, J., Yuan, L., Zhang, J., Li, H., Bai, Z., Chen, X., Zhang, W., Zhang, F. 2011. Phosphorus dynamics: from soil to plant. *Plant Physiol.* 156, 997–1005.

Shi, L., Shi, T., Broadley, M. R., White, P. J., Long, Y., Meng, J., Xu, F., and Hammond, J. P., 2013. High-throughput root phenotyping screens identify genetic loci associated with root architectural traits in Brassica napus under contrasting phosphate availabilities. *Annals of botany*. 112, 381-389.

Singh, C. P., Amberger, A., 1998. Solubilization of rock phosphate by humic and fulvic acids extracted from straw compost. *Agrochimical.* 41, 221–228.

Singh, J. S., Pandey, V. C., Singh, D. P., 2011. Efficient soil microorganisms: A new dimension for sustainable agriculture and environmental development. *Agriculture, Ecosystems & Environment*, 140, 339–353.

Stamford, N. P., Santos, P. R., Moura, A. M. M. F., Santos, C. E. R. S., Freitas, A. D. S., 2003. Biofertilizer with natural phosphate, sulphur and Acidithiobacillus in a soil with low available-p. *Sci. Agricola.* 60, 767–773.

Stamford, N. P., Santos, P. R., Santos, C. E. S., Freitas, A. D. S., Dias, S. H. L. and Lira Jr., M. A. 2007. Agronomic effectiveness of biofertilizers with phosphate rock, sulfur and Acidithiobacillus for yam bean grown on a Brazilian tableland acidic soil. *Bioresource Technol.* 98, 1311–1318.

Suresh, K. D., G. Sneh, K. K. Krishn and C. M. Mool, 2004. Microbial biomass carbon and microbial activities of soils receiving chemical fertilizers and organic amendments. *Archives Agron. Soil Sci.,* 50, 641– 647.

Tarraf, W. Ruta, C., Tagarelli, A., De Cillis, F., De Mastro, G., 2017. Influence of arbuscular mycorrhizae on plant growth, essential oil production and phosphorus uptake of Salvia officinalis L. *Industrial Crops & Products,* 102, 144–153.

Tilman, D., Cassman, K. G., Matson, P. A., Naylor, R., Polasky, S., 2002. Agricultural sustainability and intensive production practices. *Nature.* 418, 671–677.

Toyota, K., Kuninaga, S., 2006. Comparison of soil microbial community between soils amended with or without farmyard manure. *Applied Soil Ecology,* 33, 39–48.

Trabelsi, D., Cherni, A., Ben Zineb, A., Dhane, S. F. and Mhamdi R., 2017. Fertilization of Phaseolus vulgaris with the Tunisian rock phosphate affects richness and structure of rhizosphere bacterial communities. *Applied Soil Ecology,* 114, 1–8.

Valença, A. W., Bakeb, A., Brouwerb, I. D., Giller, K. E., 2017. Agronomic bioforti fi cation of crops to fi ght hidden hunger in sub-Saharan Africa, *Global Food Security.* 12, 8–14.

Wang, B., Li, J., Ren, Y., Xin, J., Hao, X., Ma, Y., Ma, X., 2015. Validation of a soil phosphorus accumulation model in the wheat–maize rotation production areas of China. *Field Crops Research,* 178, 42–48.

Weigel, H.-J. & Manderscheid, R., 2012. Crop growth responses to free air CO2 enrichment and nitrogen fertilization: Rotating barley, ryegrass, sugar beet and wheat. *European Journal of Agronomy,* 43, 97–107.

Wong, J. W. C., K. K. Ma, K. M. Fang and C. Cheung, 1999. Utilization of manure compost for organic farming in Hong Kong. *Bio-resource Technol.* 67, 43-46.

Wyngaard, N., M. L., Cabrera, K. A., Jarosch, 2016. Soil Biology & Biochemistry Phosphorus in the coarse soil fraction is related to soil organic phosphorus mineralization measured by isotopic dilution. *Soil Biology and Biochemistry,* 96, 107–118.

Xiaozhu, Y., Zhuang, L. I., Cungang, C., 2017. Effect of Conservation Tillage Practices on Soil Phosphorus Nutrition in an. *Horticultural Plant Journal,* 6, 331–337.

Xie, L., M. Liu, B. Ni, X. Zhang and Y. Wang. 2011. Slow-release nitrogen boron fertilizer from a functional superabsorbent formulation based on wheat straw and attapulgite. *Chemical Engineering Journal.* 11, 1-30.

Xie, J., Zhang, X. Y., Xu, Z. W., Yuan, G. F., Tang, X. Z., Sun, X. M., Ballantine, D. J., 2014. Total Phosphorus concentrations in surface water of typical agro- and forest ecosystems in China, 2004–2010. *Front. Environ. Sci. Eng. China.* 8, 561–569.

Zaheer, M. S., Raza, M. A. S., Saleem, M. F., Erinle, K. O., Iqbal, R. and Ahmad, S., 2019. Effect of rhizobacteria and cytokinins application on wheat growth and yield under normal vs drought conditions. *Communications in Soil Science and Plant Analysis,* 50(20), pp. 2521-2533.

Zhang, F. S., Shen, J. B., Zhang, J. L., Zuo, Y. M., Li, L. and Chen, X. P. 2010. Rhizosphere processes and management for improving nutrient use efficiency and crop productivity: implications for China. In: DL Sparks, and *Advances in agronomy,* vol. 107. San Diego: Academic Press. pp. 1–32.

Zhu, J., Li, M. and Whelan, M., 2018. Phosphorus activators contribute to legacy phosphorus availability in agricultural soils: A review. *Science of the Total Environment,* 612, pp. 522-537.

Zheng, Y., Q. Y. Xue, L. L. Xu, Q. Xul, S. Lu, C. Gu, J. H. Guo. 2011. A screening strategy of fungal biocontrol agents towards Verticillium wilt of cotton. *Biological Control.* 56, 209–216.

Ziadi, N., J. K. Whalen, A. J. Messiga, C. Morel. 2013. Assessment and Modeling of Soil Available Phosphorus in Sustainable Cropping Systems. *Advances in Agronomy,* Vol. 122.

Zornoza, R., Gómez-Garrido, M., Martínez-Martínez, S., Gómez-López, M. D., Faz, F., 2017. Science of the Total Environment Bioaugmentaton in Technosols created in abandoned pyritic tailings can contribute to enhance soil C sequestration and plant colonization. *Science of the Total Environment,* 593–594, 357–367.

Chapter 5

The Environmental Significance of Biotechnologically Treated Insoluble Phosphates and P-Solubilizing Microorganisms

Maria Vassileva[1,*]
Eligio Malusa[2]
Vanessa Martos[3]
Luis F. García del Moral[3]
Stefano Mocali[4]
Loredana Canfora[4]
Giacomo di Benedetto[5]
Aspasia Lykoudi[6]
Pedro Cartujo[7]
and Nikolay Vassilev[1,8]

[1]Department of Chemical Engineering, University of Granada, Spain
[2]The National Institute of Horticulture Research, Poland
[3]Department of Plant Physiology, University of Granada, Spain
[4]Council for Agricultural Research and Economics, Research Centre for Agriculture and Environment, Italy
[5]Enginlife, Torino, Italy
[6]Orfanos Estate Winery, Patras, Greece
[7]Department of Electronics and Computer Technology, University of Granada, Spain
[8]Institute of Biotechnology, University of Granada, Spain

---

[*] Corresponding Author's Email: mvass82@yahoo.com.

In: Biofertilizers
Editor: Philip L. Bevis
ISBN: 979-8-89113-082-1
© 2023 Nova Science Publishers, Inc.

Abstract

One of the attractive alternatives to chemical production of phosphate fertilizers is solubilization of insoluble phosphates by microorganisms. Experimentally, the microbial solubilization is based on submerged and solid-state fermentation processes using chemically defined media or agro-wastes derived from food processing and agro-industries, and organic acid producing microorganisms. The effect of the resulting fermentation products on plant growth and/or introduced into desertified and contaminated soil systems underlines the potential of these biotechnological products for application in bioremediation and environmentally sound programs and strategies. In this work, the multifaceted environmental impact of microbially treated insoluble inorganic natural sources is described. In all experimental schemes analyzed here, elements of sustainable development and circular economy in the field of agriculture could be found. In addition, novel technologies such as AI-based techniques are discussed.

Keywords: insoluble phosphates, fermentation processes, biotechnological P-solubilization, plant growth, bioremeduiation, artificial intelligence

Introduction

Although phosphorus (P) is quite abundant in many soils, it is one of the major nutrients limiting plant growth. P is added to soil in the form of phosphate fertilizers, but the overall P use efficiency is low because although plants utilize a fraction of soluble P, the rest rapidly forms insoluble complexes with soil constituents (Runge-Metzger, 1995). Therefore, frequent application of soluble forms of inorganic P is needed, well above what would be necessary under ideal conditions. Even under adequate P fertilization, only 20% or less of that applied is acquired by the first year's plant growth (Russell, 1973). As the capacity of soil to bind P is limited, this means that many soils receive P in excess of crop requirements, which often results in its leaching to the groundwater thus causing eutrophication of natural water reservoirs (Del Campillo et al., 1999).

Traditional P fertilizer production is based on chemical processing of insoluble mineral high-grade rock phosphate. What is rock phosphate? The major parts of the phosphate minerals in phosphate rocks are in fact species of apatite, what is calcium phosphate in varying combinations with quartz,

The Environmental Significance of Biotechnologically ... 119

calcite, dolomite, clay and iron oxide. Phosphate rock often contains carbonates and alkali compounds. Phosphate content in mined phosphate rocks can range from over 40% (high-grade) to below 5% (low-grade). The rock phosphate is further processed to remove the bulk of the contained impurities thus improving the rock phosphate quality. Consequently, the resulting concentrate contains improved apatite content and characteristics. The content of P_2O_5 after beneficiation of phosphate rock ranges from 26% to about 34% and up to as much as 42%. Phosphate fertilizer production is an energy intensive treatment of rock phosphate with sulfuric acid at high temperature (Goldstein, 2000). This process is environmentally undesirable, not least because of the release of contaminants into the main product, gas streams and by-products. It should be noted that lower concentration of phosphate in rock phosphate and lower quality deposit generate more waste materials, and, on the other hand, more energy and chemicals are required per tonne of useful phosphate produced (Vance, 2001).

Three facts should be taken into account when assessing the current situation in this field. One of the main reasons for concern in both phosphate rock-mining and P-fertilizer industries is that by some calculations inexpensive, high-grade rock phosphate reserves could be exhausted in as little as 60 to 80 years as the phosphate bearing ore is a finite non-renewable resource (Runge-Metzger, 1995). P fertilizer use increased 4- to 5-fold during the last 50 years and is projected to increase in the first three decades of 21st century by 20 Tg per year. Second, there are no substitutes for phosphorus in agriculture but, on the other hand, phosphorus can be recycled mainly by use of animal manure and sewage sludge (although there are serious concerns about these alternatives). Finally, it is important to note that the cultivation soil available per capita on global level is lowering while the world population is increasing with 250 000 people every day (approximately 80 million per year). This situation will result in enhanced need for food and, consequently, phosphate demand will increase. Global consumption of processed phosphate fertilizers in 2013/2014 was 40.3 Mt P2O5 (Heffer, 2015), with an annual expected increase of 1.5-2% to reach 22–27 $TgP\ yr^{-1}$ by 2050 (Zou et al., 2022). Therefore, there is an urgent need to find novel phosphate sources and/or novel techniques of application in soil-plant systems for ensuring phosphate availability.

One of the most studied approaches is the biological treatment of low-grade rock phosphates and waste materials from the rock phosphate mining and P fertilizer production (Ait-Ouakrim et al., 2023) and particularly microbial solubilization of insoluble phosphates (Vassilev et al., 2009a). In recent years various techniques for phosphate rock solubilization have been proposed, with increasing emphasis on application of free and immobilized P-solubilizing microorganisms (Vassilev, Vassileva, 2003; Vassilev et al., 2001, 2014) where the mode of the employed submerged and solid-state fermentation processes also plays an important role (Vassileva et al., 2021). In general, the P-solubilizing activity is determined by the microbial biochemical ability to produce and release metabolites such as organic acids that, through their hydroxyl and carboxyl groups, chelate the cations (mainly calcium) bound to phosphate, the latter being converted into soluble forms (Kpomblekou, Tabatabai, 1994; de Oliveira Mendes et al., 2017; Vassileva et al., 2022). It was recently shown that oxalic acid is the most effective microbially produced acid as it is able to extract 100% of P contained in different RPs. It is more efficient than sulfuric acid, as it releases more P per mol of acid applied (Mendes et al., 2020).

However, metabolizable C compounds must be applied as energy source to the microbial solubilizers to ensure their growth, organic acid production, and, simultaneously, rock phosphate solubilization (Vassilev et al., 2014). In a number of studies, it was shown that wastes from Food Industry (production of sugar, olive oil, wine, fruit juices, etc.) or from biodiesel production can be efficiently used as substrates in solubilization of low-grade (12-13% phosphate) phosphate rocks (Vassilev, Vassileva, 2003; Vassilev et al., 2014; Vassilev et al., 2016). On the other hand, we have demonstrated a number of additional advantageous multifunctional properties of the resulting fermentation products that can be summarized in four major groups (Figure 1) with a clear environmental and bioremediation potential when introduced into soil-plant systems.

The aim of this work is to critically evaluate the effectiveness of phosphate bearing, microbially treated materials in soils with a particular emphasis on their positive, ecological effect.

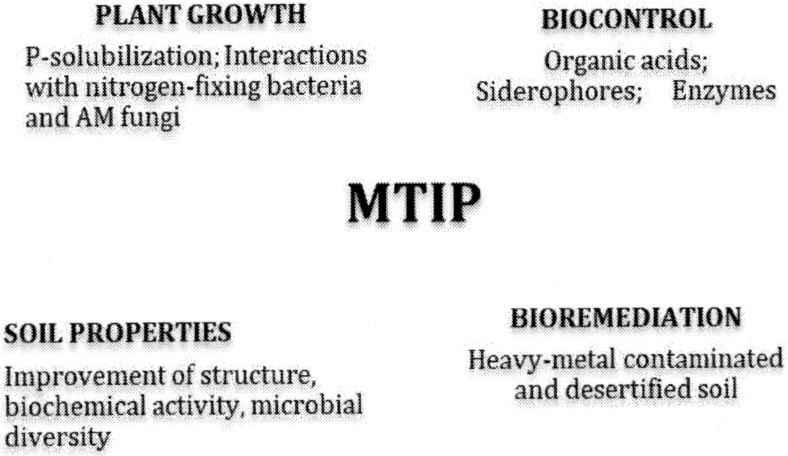

PLANT GROWTH

P-solubilization; Interactions with nitrogen-fixing bacteria and AM fungi

BIOCONTROL

Organic acids; Siderophores; Enzymes

MTIP

SOIL PROPERTIES

Improvement of structure, biochemical activity, microbial diversity

BIOREMEDIATION

Heavy-metal contaminated and desertified soil

Figure 1. Effects of Microbially Treated Inorganic Phosphates (MTIP) on plant growth, soil properties, biocontrol, and bioremediation processes.

Plant-Growth Promoting Activity of Rock Phosphate/ P-Solubilizing Microorganism Systems

The main function of the fermentation final products, resulting particularly from solid-state processes and containing mineralized organic matter, partially solubilized rock phosphate, and microbial biomass, introduced into soil-plant systems is to ensure high efficiency in enhancing plant growth (Vassilev et al., 1996a, 2006a; Vassileva et al., 1998; Mendes et al., 2015). This particular function was attributed to interrelated activities including solubilization of insoluble phosphate and interactions with other soil microorganisms such as nitrogen-fixing bacteria and arbuscular mycorrhizal fungi (AM). In fact, the main factor determining the plant growth increase is the presence of plant-available phosphate solubilized during the fermentation process. Simultaneously, the P-solubilizing microorganism interacts with the soil microflora thus creating a plant growth beneficial microbial microenvironment. This interaction was additionally supported by the organic matter, mineralized during the fermentation (Rodriguez et al., 1998; Vassileva et al., 1998; Medina et al., 2004, 2007; Vassileva et al., 2022).

The experimental studies showed that the effects of biotechnological products containing partially solubilized rock phosphate on plant growth and health were more pronounced in combination with AM (Vassilev et al., 2002;

Medina et al., 2007). In a study with isotopic ^{32}P dilution technique, it was reported that in treatments having lower specific activity values (radioactivity per amount of total P content in the plant) plants use extra ^{31}P released from otherwise unavailable P sources (Vassilev et al., 2002). The phosphate solubilizing activity of *A. niger*, used in this study as a P-solubilizing agent, could in part release P ions from the added RP and the AM (*Glomus intraradices*) external mycelium was transferring ^{31}P released from RP particles to the plant, thus inducing a lowering in the $^{32}P/^{31}P$ ratio. These results underline the particular P-fertilizing nature of the fermentation product containing solubilized P and *A. niger*, the latter being actively providing more soluble P after inoculation into soil-plant systems and continuously ensuring plant available P. Such biotechnological approach avoids P leaching and enhances the environmental importance of the fermentation final product avoiding the introduction of chemical P fertilizers.

Another question is how a product consisting of P-solubilizing microorganism/mineralized organic matter/partially solubilized rock phosphate affects the development and functionality of AM fungi and other soil microorganisms? An increased AM fungal growth and activity was found in the presence of microbially-treated organic wastes and rock phosphate introduced into plant-soil systems in compartmentalized growth units (Medina et al., 2007). It was suggested that modification of soil microbial structure and production of exudates by the P-solubilizing microorganism could explain the hyphal growth increase of three tested AM fungi. On the other hand, it is well known that AM fungi affect microbial community in both direct and indirect ways, which finally impact the overall plant development and yield (Turnau, and Haselwandter, 2002, Akyol et al., 2019). Similar positive effects were observed in experiments with nitrogen-fixing bacteria which activity was stimulated in the presence of solubilized rock phosphate and P-solubilizing microorganisms (Vassilev et al., 1996a). However, we should improve our knowledge on processes and their biological mechanisms that affect the efficacy of plant beneficial microorganisms in crop systems including plant genotypic effects, rhizosphere and soil microbiome composition, formulation, and production of bioinocula and commercial agronomic practices (Vassileva et al., 2020a; Vassilev et al., 2021). The environmental effect of application of microbially solubilized insoluble phosphate employing agro-wastes in conditions of solid-state and submerged fermentation processes was clear, particularly in typical Mediterranean soils, characterized by low organic matter, poor soil structure, low water-holding capacity, and low plant available P (Medina, Azcon, 2010). It should be noted that in the majority of cases the

The Environmental Significance of Biotechnologically ... 123

final biotechnological products with plant beneficial properties are solid products, derived from solid-state fermentations based on substrates such as sugar cane bagasse, sugar beet wastes, olive cake, and dry olive cake (Vassilev, Mendes, 2018). Similarly, liquids produced after submerged fermentation processes aimed at P-microbial-solubilization were equally efficient in promoting plant growth in soil with the same characteristics (Vassilev et al., 1998; Ceretti et al., 2004; Vassilev et al., 2017; Mendes et al., 2017). For example, the beneficial effects of *Aspergillus niger* treated olive mill wastewater (OMW) enriched with RP were highest after a repeated-batch fermentation process. The treated wheat (*Triticum durum* Desf.) showed an increase in seed biomass, spike number, and kernel weight and the harvest index was highest (0.49 ± 0.04) (Ceretti et al., 2004). Another, similar biofertilizer preparation consisted in partial incineration of solid-state fermentation derived mixture of sugar cane bagasse, fungal mycelium, and solubilized RP at 350°C or 500°C to reduce its volume and, consequently, increase P concentration (Mendes et al., 2015). The incinerated product was further applied to the soil and increased the growth and P uptake by common bean plants. The environmental advantages of these type of formulations are that they ensure a yield comparable to treatments with triple superphosphate (on a dry mass basis) and the overall proposed process is a promising energy-saving alternative for the management of chemical-P-fertilization since it enables the utilization of low-soluble and low-grade RPs and relies on the use of inexpensive substrates.

Biocontrol Activity of Rock Phosphate/Microorganism System

The beneficial environmental effect of microbially treated agro-industrial wastes and partially solubilized RP was further extended to soil-plant systems infected with fungal pathogens. The biocontrol potential of our biotechnological products was announced using agro-wastes as substrates (Vassilev et al., 2005) and proved later in a series of studies where the test phytopathogen was *Fusarium oxysporum* (Vassilev et al., 2008a). A number of studies were performed in fermentation conditions to evaluate the biocontrol activity of the microbial metabolites. Here, several groups of metabolites were proved to exert biocontrol functions: organic acids, siderophores, and enzymes such as quitinase, phytase and manganese peroxidase (Vassilev et al., 2007; Vassilev et al., 2008a, b; Vassilev et al., 2009b).

In general, organic acids are considered toxic to wide range of plant pathogens (Jain et al., 2015; Jang et al., 2016; Jayaraj et al., 2010). Among organic acids, citric, oxalic, fusaric, malonic, tartaric, propionic and lactic acids are most important as some of them are reported to enhance the hydrolytic activity of chitinases and proteases in the synergic biocontrol actions (Bidochka and Khachatourians, 1991). In P-solubilizing studies with filamentous fungi tested in submerged or solid-state fermentatios, we found citric (Vassiev et al., 1995; Mendes et al., 2013), gluconic (Vassilev et al., 1996b; Fenice et al. 2000; Mendes et al., 2013), oxalic (Mendes et al., 2013; Mendes et al., 2020), and itaconic acid (Vassilev et al., 2012; Vassileva et al., 2020b) individually or in various combinations. Apart of their high acidity, the biocontrol mechanism of organic acids most likely include induced systemic resistance in plants against diseases caused by fungi, bacteria, and viruses thus increasing defence-related enzyme activities and production of secondary metabolites, such as phenolics (Martinez-Espla et al., 2014).

It is known that in the field of agriculture, siderophores promote the growth of plant species and increase their yield by enhancing the Fe uptake to plants. Different types of siderophores are potent competitors for Fe thus reducing the Fe availability for the phytopathogens (Kloepper at al., 1980). All fungal microorganisms used in the P-solubilizing experiments (*A. niger, A. terreus, Penicillium janthinelum, Phanerochaete chrysosporium*) were also characterized by a high siderophore production demonstrated in plate assays (Vassilev et al., 2008a, b; 2009c; 2012) within their multifunctional characteristics (Vassileva et al., 2010). The importance of siderophores as biocontrol agents should be accepted in combination with their role in bioremediation strategies, particularly in detoxification of heavy-metal and radionuclides contaminated soils (Saha et al., 2016).

Extracellular enzymes with hydrolytic activity produced by soil microorganisms ensure their hyperparasitic action when attacking cell walls of phytopathogens (Chemin and Chet, 2002). Hydrolytic enzymes such as chitinases, Mn-peroxidases, laminarinases, and proteases are reported to supress fungal pathogens. In fermentation broth of microbially based P-solubilizing experiments we found chitinase (Vassilev et al., 2008b), Mn-peroxidse (Vassilev et al., 2009a), and glucose-oxidase (Fenice et al., 2000) what explains the biocontrol function of the tested microorganisms. It is important to note that the production activity of the above-mentioned metabolites by the phosphate solubilizing microorganisms was higher in the presence of rock phosphate in the medium (Vassilev et al., 2008a).

The general conclusion of these experiments is that during the fermentation processes the phosphate released continuously from the phosphate rock benefits the biochemical microbial activity and particularly that suppressed by high concentration of soluble phosphate. Similar, high metabolic activity of the P-solubilizing microorganisms was proved in different soil-plant systems (Vassilev et al., 2006b). On the other hand, bearing in mind that micronutrients found in natural phosphates such as zinc and cooper, are also known to suppress pathogens (Duffy, Defago 1999), we can explain the complex biocontrol properties described in these studies. In conclusion, the multiple interactions between all elements of the biotechnological final products including microbe-microbe and microbe-plant beneficial actions within a given soil-plant system, the overall improvement of its health without chemical products determines the environmental importance of the strategy presented here.

Effects of Rock Phosphate/Microorganism Systems on Soil Properties

During the course of studies with microbially treated agro-wastes and RP by fungal microorganisms, we have demonstrated the multiple beneficial effects of the biotechnological products on a number of soil properties such as soil structure, soil enzyme activity, and soil microbial community. The fermentation mixture of mineralized organic matter and solubilized RP raised the soil pH and increased soil water-soluble carbon under either watering or drying conditions. The beneficial effect of microbially treated organic matter and rock phosphate related to stabilization of soil aggregates was found in semiarid soil (Carrasco et al., 2009). Particularly after soil drying, aggregate stability was 66% higher than the control soil. Soil surface structure stabilization could reduce erosion and protect soil against degradation particularly when dealing with fragile, semiarid soils, which are exposed to a high risk of water erosion. Soil structure has a prevailing role in soil infiltration and biogeochemical processes. Therefore, improved soil structure means increased water retention, nutrient uptake, drainage, aeration and root growth.

On the other hand, the increase of enzymatic activities in soils resulted in an increase in the availability of nutrients to plants (including phosphorus derived from the rock phosphate), which, in turn, have a positive influence on soil fertility (Medina et al., 2004). Enzyme dehydrogenase and phosphatase

126 Maria Vassileva, Eligio Malusa, Vanessa Martos et al.

activities (indexes of microbial activity and phosphorus mineralization, respectively) were registered at maximal level when microbially treated SB/RP and mycorrhizal fungus were applied, which is an indication of their mutual effect on nutrient cycling and energy flow. In addition, the biotechnological product supplemented with rock phosphate was found to increase Indole-Acetic-Acid production in rhizosphere soil.

It was evidenced in a specific study that microbially treated agro-wastes and rock phosphate amendment is a suitable tool for increasing and changing the bacterial community in rhizosphere (Azcon et al., 2009). In this work, the bacterial-community profiles were generated from DGGE of the amplified soil DNA. Soil microbial properties such as biodiversity and dominance index increased by the application of the treated agro-waste and concomitantly favored the plant development. An important result was that rock phosphate fertilization and single mycorrhizal inoculation (used in parallel) similarly promoted plant biomass, but only mycorrhizal inoculation increased microbial diversity in the presence of agro-waste amendment. Therefore, the application of biotechnological product with the above characteristics seems to play a decisive role in improving soil properties.

Bioremediation Effects of Rock Phosphate/Microorganisms

Experiments based on the use of phosphate-bearing materials aimed at soil rehabilitation have been oriented to reclamation of soils contaminated with heavy metals and afforestation of desertified Mediterranean sites.

Most heavy metals exist naturally in the earth's crust at trace concentrations providing trace nutrients for the optimal development and metabolic functioning of the local biota (Bodek et al. 1988). However, as a result of human activities, including intentional applications, inadequate residue disposal, accidental wastes and inappropriate use, soils contain a wide number of contaminants that vary in composition and concentration (Knaebel et al., 1994; Kavamura, Esposito, 2010)). Amongst the most widespread is heavy metal contamination as a result of industrial activities such as metal manufacture and mining industries (Zhang et al., 2005). Cadmium, lead, cobalt, copper, mercury, nickel, selenium and zinc are among the metals found in soils more frequently, presenting toxic levels thus endangering both wildlife and people. In these cases, technological measures and efforts to detoxify the soils are needed. Traditional remediation methods to immobilize metals in contaminated soils that include addition of chemicals, *in-situ* vitrification,

The Environmental Significance of Biotechnologically ... 127

excavation and confinement in special waste facilities are expensive, and although efficient, often are reported to damage or change soil characteristics, including microbial communities, and need further improvement or development of novel processes (Cunningham et al. 1995; Nejad et al., 2019; Correia et al., 2020) that at least do not affect soil system productivity.

Alternative, environmentally friendly *in situ* technologies that could solve metal contamination problems include different strategies depending on the type of the soil contaminant: phytostabilization, phytoimmobilization, phytoextraction, phytovolatilization, bioremediation, etc. (Glick, 2003). In general, bioremediation techniques are based on the use of microorganisms and enzymes that transform heavy metals into volatile or insoluble forms. It is well established that soil microorganisms are deeply involved in all biogeochemical processes including a wide range of naturally occurring processes, which govern mobility and bioavailability of metals (Gadd, 2004). Microorganisms play a very important role in phyto- and bioremediation strategies, interact with soil biota and especially with plants and pathogens, and of course, with the soil components, including metals. They are widespread in all type of soils and ecosystems, and depending on the conditions, may influence one or another pathway of a compound or element conversion (Shilev et al., 2012). Depending on the microorganisms and their metabolite products, heavy metals in soils can be mobilized or immobilized by leaching mechanisms, complexation, methylation, and sorption to biomass/exopolymers, sequestration/precipitation, respectively, increasing or not their mobility and solubility (Gadd, 2007). Another attractive treatment of soils contaminated with heavy metals is based on addition of phosphate-bearing materials. An interesting approach applied in *in-situ* remediation of heavy metal contaminated soil includes a variety of phosphate amendments such as soluble phosphate and P-bearing insoluble sources, mainly rock phosphate (Tauqeer et al., 2021). Ma et al. (1993) showed that hydroxyapatite is a very efficient metal immobilizer in laboratory conditions. On the other hand, it is well known that the-bioavailability of heavy metals in soil is affected by the presence of organic matter which forms complexes and chelates of varying stability (Venegas et al., 2016). In our experiments, the complexation of heavy metals with organic matter (sugar beet wastes and olive oil residues) in the environment additionally influenced the solubility and mobility of the studied metals. Finally, P-solubilizing microorganisms (particularly filamentous fungi and yeasts) introduced into heavy metal contaminated soil-plant systems can mobilize metals through reactions such as autotrophic and heterotrophic leaching, chelation, and methylation (Grey,

1998). Application of a plant growth-promoting biotechnological product based on mineralized organic matter, partially-solubilized phosphate, and the biomass of the microbial P-solubilizer gathers all advantages of the above described heavy metal remediation agents.

Bearing all these considerations in mind, our biotechnological scheme was employed for the first time in heavy metal (mainly Zn and Cd) contaminated soil rehabilitation activities (Medina et al., 2005, 2006, 2010). The results of these experiments showed that the combination of plant growth improvement and reduced metal translocation caused by our biotechnological products containing partially solubilized RP could be regarded as a promising strategy for remediating heavy metal contaminated soil. These effects were recently presented in details (Medina, Azcon, 2010) but here it would be important to include a new approach-the inclusion of animal bone char as the P-bearing source in the same biotechnological scheme for microbial solubilization. The advantages of animal bone char are numerous but particularly in treatment of heavy metal contaminated soil, the lack of heavy metals in its structure seems important in comparison with rock phosphate (Vassilev et al., 2009b; Vassileva et al., 2012; Vassilev et al., 2013). In rock phosphate and derived P fertilizers there are heavy metals in excess although the metal amounts depend on the nature of the mineral. In a recent study, it was reported that while Al, As, B, Be, Cd, Cr, Mo, Ni, Pb, Sb, Se, Tl, U, and Zn were higher in sedimentary than in igneous rock phosphates, the opposite was true for Co, Cu, Sn, Mn, Ti, Fe, and Sr (Kratz et al., 2016). Particularly high are the concentrations of Cd and U, which exceed the legal limits. Similarly, B and Fe are oversupplied in soils enriched with RP and P-bearing fertilizers and as their loads exceed plant uptake capacity, P-immobilization can be observed. In addition, partially solubilized biochar application can be even more advantageous than RP use, as biochar is known to form complexes with heavy metals facilitated by its large structure and functional groups (Lu et al., 2012), improve soil structure and microbial community, and enhance plant growth (Tammeorg et al., 2016). It is also important to note the possibility of application of P-solubilizing microorganisms in soil reclamation activity in stress conditions as recently reported experiments in presence of animal bonechar (Vassileva et al., 2023). Therefore, when assessing the potential of partially solubilized phosphates by microorganisms applied in bioremediation strategies or simply as fertilizers, we should also pay attention to the composition of the P-bearing materials in order to ensure an environmentally friendly practice.

The Environmental Significance of Biotechnologically ... 129

In another series of experiments, our biotechnological products were successfully tested in reclamation of a desertified Mediterranean site (Caravaca et al., 2004). The establishment of a plant cover based on the use of seedlings with an optimized microbiological and physiological status is paramount in order to carry out successful re-afforestation activities in desertified ecosystems, particularly those developed under Mediterranean environments. Drought tolerant, native shrub species have been recommended for re-establishment of functional shrub lands and recovery of desertified Mediterranean ecosystems. In general, the use of native plant species establishment is the most effective strategy for reclamation of desertified areas characterized by conditions of abiotic stresses such as water deficiency, lack of essential macronutrients, organic matter deficiency (Bashan et al., 2012). The results of our studies demonstrated the viability of applying the fungal processed agro-wastes in the presence of rock phosphate in order to improve the growth of the woody legume *Dorycnium pentaphyllum.* This could be due to an improvement in the available nutrient supply in the soil, arising from the fermented organic waste. During the course of microbial fermentation, the rock phosphate was partially solubilized thus increasing the level of bio-available P (Vassileva et al., 1998).

Our biotechnological scheme for preparation of soluble P bearing material rich in partly mineralized organic matter and fungal mycelial mass was further applied by other research groups in developing strategies for reclamation of degraded soils (Armada et al., 2015; Kohler et al., 2016 and related references herein). However, they paid more attention to bacterial/mycorrhizal inoculants without analyzing the effect of the fermenation product. The latter is, in addition to all of the above mentioned elements, formulated product of *A. niger* known as a producer of wide number of enzymes, siderophores, organic acids, etc. (Vassileva et al., 2010). At the end of the fermentation process, this mixture brings to soil a great part of the fermentation medium (Czapek-Dox medium). Therefore, it ensures an excellent environment for development of external inoculants and/or authoctonous microflora and is the main growth promoting factor.

Conclusion

In conclusion, in a wide number of experiments the positive effects of microbially treated low-grade rock phosphate and animal bonechar was proven. Their combinations with organic matter and microbial mass resulted highly effective in several types of soil-plant application demonstrating multiple ecological impacts with economic importance. In all these experimental schemes, elements of sustainable development and circular economy in the field of agriculture could be find. The main advantageous results that can be distinguished are related to the use low-grade non-renewable P source, the application of novel source of P such as animal bonechar, valorization of agro-industrial wastes applied as substrates in the biotechnological P-solubilization, the multifaceted use of the final fermentation products as biocontrol agents or in bioremediation of disturbed soils. Other research groups applying our biotechnological scheme, using the same residues and P-solubilizing microorganism confirm the results presented here. All the above issues are related to research activities carried out by the EXCALIBUR project (www.excaliburproject.eu), which aims to expand the current concept about microbiomes interactions, acknowledging their interactive network that can impact agricultural practices as well as on all living organisms within an ecosystem. Within the European research tendencies, a serious interest could be noted on the development and application of AI tools able to analyze real soil data and further predict the need of functional nutrient management including strategies based on biofertilizers and thus avoiding chemical products. Various technologies, such as the Internet of Things, machine learning, big data analytics, and automation will make easier to take and apply decisions regarding P bio-based nutrition, which will enhance its overal environmental impact (see the web page of the EC Project SUSTAINABLE—htpps://www. projectsustainable.eu. accessed on 20 May 2023).

Acknowledgments

This work was supported by the project EXCALIBUR funded from the European Union's Horizon 2020 research and innovation programme under grant agreement No. 817946 and project SUSTAINABLE under grant agreement no. 101007702.

The authors declare that they have no conflict of interest.

References

Ait-Ouakrim EH, Chakhchar A, El Modafar C, Allal Douira. Valorization of Moroccan Phosphate Sludge Through Isolation and Characterization of Phosphate Solubilizing Bacteria and Assessment of Their Growth Promotion Effect on *Phaseolus vulgaris* (2023) Waste Biomass Valor. https://doi.org/10.1007/s12649-023-02054-2.

Armada E, Barea J, Castillo L, Roldan A, Azcon R. Characterization and management of autochthonous bacterial strains from semiarid soils of Spain and their interactions with fermented agrowastes to improve drought tolerance in native shrub species. *Appl Soil Ecol* (2015) 96:306-318.

Akyol TY, Niwa R, Hirakawa H, Maruyama H, Sato, Takae Suzuki, Ayako Fukunaga, Takashi Sato, Shigenobu Yoshida, Keitaro Tawaraya, Masanori Saito, Tatsuhiro Ezawa, Shusei Sato. Impact of introduction of arbuscular mycorrhizal fungi on the root microbial community in agricultural fields. *Microbes and environments* (2019) *34*:23-32. https://doi.org/10.1264/jsme2.ME18109.

Azcon R, Medina A, Roldan A, Biro B, Vivas A Significance of treated agro-waste residue and autochthonous inoculates (arbuscular mycorrhizal fungi and *Bacillus cereus)* on bacterial community structure and phytoextraction to remediate soils contaminated with heavy metals. *Chemosphere* (2009) 75:327-334.

Bashan Y, Salazar BG, Moreno M, Lopez BR, Linderman RG. Restoration of eroded soil in the Sonoran Desert with native leguminous trees using plant growth-promoting microorganisms and limited amounts of compost and wáter. *J Environ Manag* (2012) 102:26–36

Bidochka MJ, Khachatourians GG. The implication of metabolic acids produced by *Beauveria bassiana* in pathogeneisis of the migratory grasshopper *Melanoplus sanguinipes. J Insect Pathol* (1991) 58:106-117.

Bodek I, Lyman WJ, Reehl W, Rosenblatt DH. (1988) *Environmental Inorganic Chemistry: Properties, Processes and Estimation Methods.* Pergamon Press, Inc, Elmsford, NY.

Caravaca F, Aguacil MM, Vassileva M, Diaz G, Roldan A AM fungi inoculation and addition of microbially-treated dry olive cake-enhanced afforestation of a desertified Mediterranean site. *Land Degr Develop* (2004) 15: 153-161

Carrasco L, Caravaca, Azcon, R, Rolda A. Soil acidity determines the effectiveness of an organic amendment and a native bacterium for increasing soil stabilization in semiarid mine tailings. *Chemosphere* (2009) 74:239-244.

Cereti CF, Rossini F, Federici F, Quarantino D, Vassilev N, Fenice M. Reuse of microbially-treated olive mill wastewater as fertilizer for wheat (*Triticum durum* Desf.). *Biores Technol (*2004) 91:135-140.

Chemin L, Chet I. Microbial enzymes in biocontrol of plant pathogens and pests. In: Burns RG, Dick RP (eds) *Enzymes in the environment: activity, ecology and applications.* Marcel Dekker, New York, (2002) pp 171–225.

Correia AAS, Matos MPSR, Gomes AR, Rasteiro MG. Immobilization of Heavy Metals in Contaminated Soils—Performance Assessment in Conditions Similar to a Real Scenario. *Appl Sci* (2020) *10:*7950. https://doi.org/10.3390/app10227950.

Cunningham SD, Berti R., Huang JW. Phytoremediation of contaminated soils. *Trends Biotechnol* (1995) 13:393–397.

de Oliveira MG, de Freitas ALM, Pereira OL, da Silva IR, Vassilev NB, Costa MD. Mechanisms of phosphate solubilization by fungal isolates when exposed to different P sources. *Ann Microbiol* (2014) 64:239–249.

Del Campillo SE, Van der Zee SEATM, Torrent J. Modelling long-term phosphorus leaching and changes in phosphorus fertility in excessively fertilized acid sandy soils. *Eur J Soil Sci* (1999) 50:391–399.

Duffy BK, Defago G. Trace mineral amendments in Agriculture for optimizing the biocontrol activity of plant-associated bacteria. In: P. Berthelin, M. Huang, JM. Bo Hag, F. Andreux, Editors. *Effect of mineral-organic-microorganism interactions on soil and freshwater environments.* New York: Kluwer Academic/Plenum Publishers, 1999 pp. 295-304

Fenice M, Selbman L, Federici F, Vassilev N. Solubilization of rock phosphate by encapsulated cells of *Penicillium variable* P16. *Biores Technol* (2000) 73:157-162.

Gadd G. Microbial influence on metal mobility and application for bioremediation. *Geoderma* (2004) 122:109-119.

Gadd G. 2007. Geomycology: biogeochemical transformations of rocks, minerals, metals and radionuclides by fungi, bioweathering and bioremediation. *Mycol. Res.* 111, 3-49.

Glick BR. Phytoremediation: synergistic use of plants and bacteria to clean up the environment. *Biotechnol Adv* (2003) 21:383–93.

Goldstein AH. Bioprocessing of rock phosphate ore: essential technical considerations for the development of a successful commercial technology. *IFA Technical Conference,* New Orleans, La. (2000) pp 1–21, http://goldsteinlab.alfred.edu/publications.html

Gray SN. Fungi as potential bioremediation agents in soil contaminated with heavy or radioactive metals. *Biochem Soc Trans* (1998) 26:666-670.

Heffer P. Fertilizer Outlook 2015-2019. In: *83rd IFA Annual Conference* (Ed.) (2015) M. Prud'homme Istabul (Turkey).

Jain A, Singh A, Singh S, Sarma BK, Singh HB. Biocontrol agents-mediated suppression of oxalic acid induced cell death during *Sclerotinia sclerotiorum*-pea interaction. *J Basic Microbiol* (2015) 55:6011–606.

Jang JY, Choi YH, Shin S, Kim TH, Shin KS, Park HW, Kim YH., Kim H, Choi GJ, Jang KS, Cha B, Kim IS, Myung EJ, & Kim JC. Biological Control of *Meloidogyne incognita* by *Aspergillus niger* F22 Producing Oxalic Acid. *PLoS ONE* (2016) 11:e0156230. doi:10.1371/journal.pone.0156230.

Jayaraj J, Bhuvaneswari R, Rabindran R, Muthukrishnan, S, Velazhahan R. Oxalic acid-induced resistance to *Rhizoctonia solani* in rice is associated with induction of phenolics, peroxidase and pathogenesis-related proteins. *J Plant Interact* (2010) 5:147–157.

Kavamura VN, Esposito E. Biotechnological strategies applied to the decontamination of soils polluted with heavy metals. *Biotechnol Adv* (2010) 28:61-69.

Kloepper JW, Leong J, Teintze M, Schirot N. Enhanced plant growth by siderophores produced by plant growth promoting rhizobacteria. *Nature* (1980) 286:885-886.

The Environmental Significance of Biotechnologically ... 133

Knaebel DB, Federle TW, McAvoy DC, Vestal R. Effect of mineral and organic soil constituents on microbial mineralization of organic compounds in a natural soil. *Appl Environ Microbiol* (1994) 60:4500–45088.

Kohler J, Caravaca F, Azcon, Diaz, G, Roldan A. Stability of the microbial community composition and function in a semiarid mine soil fos assessing phytomanagement practices based on mycorrhizal inoculation and amendment addition. *J Environ Manag* (2016) 169:236-246.

Kpomblekou K, Tabatabai MA. Effect of organic acids on release of phosphorus from phosphate rocks. *Soil Sci* (1994) 158:442–453.

Kratz S, Schick J, Schnug E. Trace elements in rock phosphates and P containing mineral and organo-mineral fertilizers sold in Germany. *Sci. Total Environ* (2016) 542:1013-1019.

Lu H, Zhang W, Yang Y, Huang X, Wang S, Qiu R. Relative distribution of Pb^{2+} sorption mechanisms by sludgeederived biochar. *Water Res* (2012) 46:854-862.

Ma QY, Traina SJ, Logan TJ, Ryan JA. *In situ* lead immobilization by apatite. *Environ Sci Technol* (1993) 27:1803-1810.

Martinez-Espla A, Zapata PJ, Valero D, Garcia-Viguera C, Castillo SC, Serrano M. Preharvest application of oxalic acid increased fruit size, bioactive compounds, and antioxidant capacity in sweet cherry cultivars (*Prunus avium* L.). *J Agric Food Chem* (2014) 62:3432–3437.

Medina A, Vassilev N, Alguacil M, Roldan A, Azcon R. Increased plant growth, nutrient uptake and soil enzymatic activities in a decertified Mediterranean soil amended with treated residues and inoculated with native AM fungi and plant-growth-promoting yeast. *Soil Sci* 2004:169:260-270.

Medina A, Vassilev N, Barea JM, Azcon R. Application of *Aspergillus-treated* agrowaste residue and *Glomus mosseae* for improving growth and nutrition of *Trifolium repens* in a Cd-contaminated soil. *J Biotechnol* (2005) 116:369-378.

Medina A, Vassileva M, Bare JM, Azcon R The growth-enhancement of clover by *Aspergiilus*-treated sugar beet waste and *Glomus mosseae* inoculation in Zn contaminated soil. *Appl Soil Ecol* (2006) 33:87-98.

Medina A, Jakobsen I, Vassilev N, Azcón R, Larsen J. Fermentation of sugar beet waste by *Aspergillus niger* facilitates growth and P uptake of external mycelium of mixed populations of arbuscular mycorrhizal fungi. *Soil Biol Biochem* 2007 39:485-492.

Medina A, Azcon R. Effectiveness of the application of arbuscular mycorrhizal fungi and organic amendements to improve soil quality and plant performance under stress conditions. *J Soil Sci Plant* Nutr (2010) 10:354-372.

Medina A, Vassilev N, Azcon R. The interactive effect of an AM fungus and an organic amendment with regard to improving inoculum potential and the growth and nutrition of *Trifolium repens* in Cd-contaminated soils. *Appl Soil Ecol* (2010) 44:181-189.

Mendes G, Vassilev N, Araújo VH, da Silva IR, Júnior JI, Costa M. Inhibition of fungal phosphate solubilization by fluoride released from rock phosphate. *Appl Environ Microbiol* (2013) doi: 10.1128/AEM.01487-13.

Mendes G, Silva NMRM, Anastacio TC, Vassilev NB. Optimization of *Aspergillus niger* phosphate solubilization in solid state fermentation and use of the resulting product as a P fertilizer. *Microb Biotechnol* Thematic Issue: *Fungal Biotechnol* 2015 8:930-939.

Mendes G, Galvez A, Vassileva M, Vassilev N. Fermentation liquid containing microbially solubilized P significantly improved plant growth and P uptake in both soil and soilless experiments. *Appl Soil Ecol* 2017 117-118:208-211, doi 10.1016/j.apsoil.2017.05.008.

Mendes GO, Murta H, Valadares RV, da Silveira WB, da Silva IR, Costa MD. Oxalic acid is more efficient than sulfuric acid for rock phosphate solubilization. *Miner Eng* (2020) 155:106458.

Nejad ZD, Jung MC, Kim KH Remediation of soils contaminated with heavy metals with an emphasis on immobilization technology. *Environ Geochem Health* (2019) 40:927–953.

Rodriguez R, Vassilev N, Azcon R. Increases of growth and nutrient uptake of alfalfa grown in soil amended with microbially-trated sugar beet waste. *Appl Soil Ecol* (1998) 330:1-7.

Runge-Metzger A. Closing the cycle: obstacles to efficient P management for improved global security. In: Tiessen H, ed. Phosphorus in the global environment (1995) Chichester, UK: John Wiley and Sons Ltd, 27–42.

Russell DW. Soil conditions and plant growth (1973) New York, NY, USA: Longman Group Ltd.

Saha M, Sarkar S, Sarkar B, Sharma B, Bhattacharjee S, Tribedi P. Microbial siderophores and their potential applications: a review. *Environ Sci Pollut Res* (2016) 23:3984–3999. https://doi.org/10.1007/s11356-015-4294-0.

Shilev S, Naydenov M, Prieto MS, Vassilev N, Sancho ED PGPR as inoculants in management of lands contaminated with trace elements. In: *Bacteria in Agrobiology*: *Stress Management*, DK Maheshwari (ed.), (2012) pp. 259-277. Springer-Verlag Berlin Heidelberg.

Simons AM, Milkiyas A, Garrick B, Bourcard N. Indigenous bone fertilizer for growth and food security: A local solution to a global challenge. *Food Policy* (2023) 114:102396 https://doi.org/10.1016/j.foodpol.2022.102396.

Tammeorg P, Bastos AC, Jeffery S, Rees F, Kern J, Graber ER, Ventura M, Kibblewhite M, Amaro A, Budai A, Cordovil CMDS, Domene X, Gardi C, Gascó G, Horák J, Kammann C, Kondrlova E, Laird D, Loureiro S, Verheijen FGA. Biochars in soils: towards the required level of scientific understanding. *J Environ. Eng Landscape Manag* (2016) doi.org/10.3846/16486897.2016.1239582.

Tauqeer HM, Maryam Fatima, Audil Rashid, Ali Khan Shahbaz, Pia Muhammad Adnan. The Current Scenario and Prospects of Immobilization Remediation Technique for the Management of Heavy Metals Contaminated Soils. In: Hasanuzzaman, M. (eds) *Approaches to the Remediation of Inorganic Pollutants*. Springer, (2021) Singapore. https://doi.org/10.1007/978-981-15-6221-1_8.

Turnau K, Haselwandter K. Arbuscular mycorrhizal fungi, an essential component of soil microflora in ecosystem restoration. In: S. Gianinazzi, H. Schüepp, J. M. Barea, K. Haselwandter (eds). *Mycorrhiza Technology in Agriculture, from Genes to Bioproducts*. Birkhauser Verlag, Basel, Switzerland (2002) pp: 137-149.

Vance CP. Symbiotic nitrogen fixation and phosphorus acquisition. Plant nutrition in a world of declining renewable resources. *Plant Physiol* 2001 127:390–397.

The Environmental Significance of Biotechnologically ... 135

Vassilev N, Baca MT, Vassileva, Franco I, Azcon R. Rock phosphate solubilization by *Aspergillus niger* grown on sugar beet wastes. *Appl Microbiol Biotechnol* (1995) 44:546-549.

Vassilev N, Franco I, Vassileva M, Azcon R. Improved plant growth with rock phosphate solubilized by *Aspergillus niger* grown on sugar beet waste. *Biores Technol* (1996a) 55:237-241.

Vassilev N, Fenice M, Federici F. Rock phosphate solubilization with gluconic acid produced by *Penicillium variable* P16. *Biotechnol Tech* (1996b) 10:684-688.

Vassilev N, Vassileva M, Fenice M, Federici F, Azcon R, Barea JM. Fertilizing effect of microbially-treated olive oil wastewater on *Trifolium* plants. *Biores Technol* (1998) 66:133-137.

Vassilev N, Vassileva M, Fenice M, Federici F. Immobilized cell technology applied in solubilization of insoluble inorganic (rock) phosphates and P plant acquisition. *Biores Technol* (2001) 79: 263-271.

Vassilev N, Vassileva M, Azcon R, Barea JM. The use of ^{32}P dilution techniques to evaluate the effect of mycorrhizal inoculation on plant uptake of P from the resulting products of fermentation mixtures including agrowastes, *Aspergillus niger* and rock phosphate. In: *The use of Nuclear and Related Techniques for Evaluating the Agronomic Effectiveness of Phosphate fertilizers.* International Atomic Energy Agency Technical Document. FAO/IAEA (2002) Vienna, Austria.

Vassilev N, Vassileva M. Biotechnological solubilization of rock phosphate on media containing agro-industrial wastes. *Appl Microbiol Biotechnol* (2003) 61:435-440.

Vassilev N, Nikolaeva I, Vassileva M. Biocontrol properties of microbially treated sugar beet waste in presence of rock phosphate. *J Biotechnol* (2005) 118, Suppl S177.

Vassilev N, Medina A, Azcon R, Vassileva M. Microbial solubilization of rock phosphate on media containing agro-industrial wastes and effect of the resulting products on plant growth and P uptake. *Plant Soil* 2006a) 287:77-84.

Vassilev N, Vassileva M, Nikolaeva I. Simultaneous P-solubilizing and biocontrol activity of microorganisms: Potentials and future trends. *Appl Microbiol Biotechnol* (2006b) 71:137-144.

Vassilev N, Vassileva M, Bravo V, Fernandez-Serrano M, Nikolaeva I. Simultaneous phytase production and rock phosphate solubilisation by *Aspergillus niger* grown on dry olive wastes. *Ind Crops Prod* (2007) 26:332-336,

Vassilev N, Nikolaeva I, Jurado E, Reyes A, Fenice M, Vassileva M. Antagonistic effect of microbially-treated mixture of agro-industrial wastes and inorganic insoluble phosphate to *Fusarium* wilt disease. In: Myung-Bo Kim, Ed., *Progress in Environmental Microbiology*, Nova Publishers, USA, (2008a) pp. 223-234.

Vassilev N, Reyes A, Garcia M, Vassileva M. Production of chitinase by free and immobilized cells of *Penicillium janthinellum*. A comparison between two biotechnological schemes for inoculant formulation. *Proceedings XVI International Conference on Bioencapsulation.* Dublin. 5-6 Sep 2008. Publ by DCU-Ireland (2008b) pp P07 1-4.

Vassilev N, Requena A, Nieto L, Nikolaeva I, Vassileva M. Production of manganese peroxidase by *Phanerochaete chrysosporium* grown on medium containing agro-wastes/rock phosphate and biocontrol properties of the final product. *Ind Crops Prod* (2009a) 30, 28–32.

Vassilev N, Serrano M, Jurado E, Bravo V, Nikolaeva I, Vassileva M. Effect of microbially treated agro-wastes and simultaneously solubilized animal bone char on Zn and P uptake and growth of white clover in Zn-contaminated soil. *New Biotechnol* (2009b) 25, S312.

Vassilev N, Someus E, Serrano M, Bravo V, Garcia Roman M, Reyes A, Vassileva M. Novel approaches in phosphate-fertilizer production based on wastes derived from rock phosphate mining and food processing industry. In: *Industrial Waste: Environmental Impact, Disposal and Treatment,* Samuelson JP (ed), (2009c) Nova Publishers, USA, pp. 387-391.

Vassilev N, Medina A, Eichler-Löbermann B, Flor-Peregrín E, Vassileva M. Animal Bone Char Solubilization with Itaconic Acid Produced by Free and Immobilized *Aspergillus terreus* Grown on Glycerol-Based Medium. *Appl Biochem Biotechnol* (2012) 168:1311-1318.

Vassilev N, Martos E, Mendes G, Flor-Peregrin E, Martos V, Vassileva M. Biochar of animal origin: A sustainable solution of the high-grade rock phosphate scarcity. *J Sci Food Agric* (2013) 93:1799-1804.

Vassilev N, Mendes G, Costa M, Vassileva M. Biotechnological tools for enhancing microbial solubilization of insoluble inorganic phosphates. *Geomicrobiol J* (2014) 31:751-763.

Vassilev N, Malusà E, Reyes A, Martos V, Lopez A, Maksimovic I, Vassileva M. Potential application of glycerol in the production of plant beneficial microorganisms. *J Ind Microbiol Biotechnol* (2017) 44:735–743; doi.org/10.1007/s10295-016-1810-2

Vassilev N, Eichler-Löbermann B, Flor-Peregrin E, Martos V, Reyes A, Vassileva M. Production of a potential liquid plant bio-stimulant by immobilized *Piriformospora indica* in repeated-batch fermentation process. *AMB Express* (2017) 7(1):106; doi 10.1186/s13568-017-0408-z.

Vassilev N, Mendes G. Solid-State Fermentation and Plant-Beneficial Microorganisms,. In: *Current Developments in Biotechnology and Bioengineering*, Eds. A Pandey, Ch Larroche, CR Soccol, Elsevier (2018) pp. 435-450, http://dx.doi.org/10.1016/B978-0-444-63990-5.00019-0.

Vassilev N, Malusà E, Neri D, Xu X. Editorial: Plant Root Interaction With Associated Microbiomes to Improve Plant Resiliency and Crop Biodiversity. *Front Plant Sci* (2021) 12:715676. doi: 10.3389/fpls.2021.715676.

Vassileva M, Medina A, Reyes A, Martos V, Vassilev N. Remediation of Heavy Metal Contaminated Soils by Phosphate-Bearing Biotechnological Products. In: *Bioremediation: Biotechnology, Engineering and Environmental Management*: ISBN: 978-1-61122-730-7. (2012) Nova Publishers, USA, 465-474.

Vassileva M, Serrano M, Bravo V, Jurado E, Nikolaeva I, Martos V, Vassilev N. Multifunctional properties of phosphate-solubilizing microorganisms grown on agro-industrial wastes in fermentation and soil conditions. *Appl Microbiol Biotechnol* (2010) 85:1287-1299.

The Environmental Significance of Biotechnologically ...   137

Vassileva M, Vassilev N, Azcon R. Rock phosphate solubilization by *Aspergillus niger* grown on olive cake and its further introduction into soil-plant system. *World J Microbiol Biotechnol* (1998) 14:281-284.

Vassileva M, Flor-Peregrin E, Malusá E, Vassilev N. Towards Better Understanding of the Interactions and Efficient Application of Plant Beneficial Prebiotics, Probiotics, Postbiotics and Synbiotics. *Front Plant Sci* (2020a) 11:1068. doi: 10.3389/fpls.2020.01068

Vassileva M, Malusà E, Lobermann B, Vassilev N. *Aspergillus terreus*: From soil to Industry and back. *Microorganisms* (2020b) 8, 1655; doi:10.3390/micro organisms8111655.

Vassileva M, Malusà E, Sas-Paszt L, Trzcinski P, Galvez A, Flor-Peregrin E, Shilev S, Canfora L, Mocali S, Vassilev N. Fermentation Strategies to Improve Soil Bio-Inoculant Production and Quality. *Microorganisms* (2021) 9, 1254. https://doi.org/ 10.3390/microorganisms9061254.

Vassileva M, Mendes GdO, Deriu M., Benedetto Gd, Flor-Peregrin E, Mocali S, Martos V, Vassilev N Fungi, P-Solubilization, and Plant Nutrition. *Microorganisms* (2022) 10, 1716. https://doi.org/10.3390/microorganisms10091716.

Vassileva M, Martos V, del Moral LFG, Vassilev N. Effect of the Mode of Fermentation on the Behavior of *Penicillium bilaiae* in Conditions of Abiotic Stress. *Microorganisms*, (2023) 11, 1064. https://doi.org/10.3390/microorganisms 11041064.

Venegas A, Rigol A, Vidal M. Changes in heavy metal extractability from contaminated soils remediated with organic waste or biochar. *Geoderma* (2016) 279:132-140.

Zhang GL, Yang FG, Zhao YG, Zhao WJ, Yang JL, Gong ZT Historical change of heavy metals in urban soils of Nanjing, China during the past 20 centuries. *Environ Int* (2005) 31:913–919.

Zou T, Zhang X, Davidson EA. Global trends of cropland phosphorus use and sustainability challenges *Nature* (2022) 611:81–87. https://doi.org/10.1038/s41586-022-05220-z.

Chapter 6

A Sustainable Approach to Increase the Bioavailability of Iron in Plants: The Potential of Iron-Solubilizing Microbes

Bahman Khoshru[1], PhD
and Mohammad Reza Sarikhani[*], PhD

[1]Soil and Water Research Institute, Agricultural Research,
Education and Extension Organization (AREEO), Karaj, Iran
[2]Department of Soil Science, Faculty of Agriculture, University of Tabriz, Iran

Abstract

Iron (Fe) is an essential nutrient required for plant growth and development, and its availability in the soil is crucial for meeting the nutritional needs of society. However, Fe availability in soil is affected by various factors, including soil pH, organic matter content, and the form of Fe present in the soil. Fe deficiency in plants is a widespread problem that affects crop productivity and quality. Traditional practices such as chemical fertilizers and amendments have been used to increase Fe availability in soil, but the negative effects of chemical fertilizers on the environment and human health are obvious. Therefore, sustainable agriculture practices that rely on the natural abilities of microorganisms to solubilize Fe in the soil are gaining more attention. This chapter discusses the importance of Fe in soil and plants, the factors affecting its solubility and deficiency in the soil, and the potential of microbial iron solubilizers as a sustainable approach to enhancing Fe availability in soil and promoting plant growth. Microorganisms such as bacteria and fungi have the ability to solubilize Fe in the soil through various mechanisms,

---

[*] Corresponding Author's Email: rsarikhani@yahoo.com

In: Biofertilizers
Editor: Philip L. Bevis
ISBN: 979-8-89113-082-1
© 2023 Nova Science Publishers, Inc.

including acidification, chelation, and reduction. The use of microbial solubilizers can improve Fe availability in the soil, enhance plant growth, and reduce the need for chemical fertilizers. Therefore, use of microbial iron solubilizers can be a viable and cost-effective approach to enhance Fe availability in soil and promote sustainable agriculture. Further research is needed to explore the potential of different microbial strains and their interactions with plants and other microorganisms in the soil. This chapter highlights the importance of Fe availability in soil and plants and the potential of microbial iron solubilizers as a sustainable approach to address Fe deficiency in soil and promote sustainable agriculture.

Keywords: Fe solubilizing microbes, bacteria, fungi, organic acids, siderophore

Introduction

Iron (Fe) is a crucial micronutrient for plant growth, as it plays a vital role in the production of chlorophyll and other important enzymes and proteins (Rout and Sahoo, 2015). However, the availability of Fe in the soil can be limited, as it is often present in insoluble forms, such as Fe oxides and hydroxides (Colombo et al., 2015). In order to absorb Fe by plants, it must be converted into a soluble form, which can then be taken up by their roots. This conversion process is known as Fe solubilization and can be carried out by various microorganisms, including bacteria and fungi (Mishra et al., 2016). One important factor that influences the availability of Fe in soil is pH. At low pH levels, Fe is more soluble and therefore more available to plants. However, at high pH levels, Fe can become tied up in insoluble forms, which reduces its availability to plants (Colombo et al., 2014). This is why it is important to maintain appropriate soil pH levels for optimal plant growth. Another important factor that affects the Fe cycle in the soil is the oxidation-reduction potential (Zhang et al., 2012). This refers to the tendency of soil to either gain or lose electrons, which can influence the solubility of Fe. In anaerobic conditions, such as waterlogged soil or paddy soil, Fe can become reduced and more soluble. Conversely, in aerobic conditions, such as well-drained soil, Fe can become oxidized and less soluble (Pearsall and Mortimer, 1939).

Microorganisms play a key role in the Fe cycle in soil. Some bacteria, such as *Pseudomonas* and *Bacillus* species, are known to produce siderophores, which are molecules that chelate Fe and increase its solubility (Crowley, et al., 1991, Khoshru et al., 2022). Other bacteria, such as

Rhizobium and *Bradyrhizobium* species, can form symbiotic relationships with legume plants, in which they provide the plant with biologically available Fe in exchange for nutrients from the plant (Mabrouk et al., 2018). Fungi can also play a role in solubilizing Fe, particularly in forest soils where they contribute to the breakdown of organic matter (Rashid et al., 2016). In addition to microbial activity, physical and chemical processes can also influence the Fe cycle in soil. For example, Fe can be lost from the soil through erosion or leaching, particularly in areas with high rainfall or poor soil management practices (Nair and Nair, 2019). Conversely, Fe can accumulate in the soil over time through weathering of parent materials or deposition from the atmosphere. Maintaining an adequate supply of Fe in the soil is essential for maximizing plant growth and productivity. This can be achieved through the use of Fe-containing fertilizers or by promoting the activity of microorganisms that can solubilize Fe (Khoshru et al., 2020b). Understanding the Fe cycle in the soil is also important for understanding the role of soil ecosystems in global biogeochemical cycles, as Fe is an important nutrient for many organisms in aquatic and terrestrial environments. Therefore, the Fe cycle in the soil is a complex and dynamic process that is influenced by a variety of factors, and it is essential for maintaining soil fertility and ecosystem health (Strawn et al., 2020).

Fe deficiency in soils is a common problem worldwide, particularly in alkaline or poorly drained soils (Vose, 1982). The exact prevalence of Fe deficiency in soils is difficult to determine, as it can vary widely depending on factors such as geography, climate, soil type, and agricultural practices. However, there are some statistics available that provide insight into the scope of the problem: a) According to the Food and Agriculture Organization (FAO), Fe deficiency is the most common micronutrient deficiency in the world, affecting an estimated two billion people globally (Smith et al., 2017). b) Fe deficiency is particularly prevalent in developing countries, where diets are often low in Fe and soil fertility is poor. In sub-Saharan Africa, for example, an estimated 40% of the population is affected by Fe deficiency anemia (Mwangi et al., 2021). c) Fe deficiency in soils can also have a significant impact on crop yields and food security. According to the FAO reports, Fe deficiency is a major constraint to agricultural productivity in many parts of the world, particularly in sub-Saharan Africa and South Asia (Gregory et al., 2017). d) A new study published in the Journal Nature Plants (2019) found that Fe deficiency affects approximately 30% of global agricultural lands and that this problem is likely to worsen in the coming decades due to changing climate patterns and land-use practices (Arora et al., 2018). e) In some regions, such as the Pacific Northwest in the United States, Fe deficiency can be a significant problem in high-pH soils. According to a study published in the Journal of Plant Nutrition in 2015, up to 70% of wheat fields in the region were found to be deficient in Fe (Thapa et al., 2021).

Generally, Fe deficiency in soils is a significant global problem that affects both human nutrition and agricultural productivity. Addressing this problem will require a complex approach that includes improving soil management practices, promoting the use of Fe-containing fertilizers, and developing crops that are more resilient to Fe deficiency (Graham, 2008).

Fe in Soil and Its Availability

In soil, iron exists in various forms, including crystalline and amorphous iron oxides, hydroxides, and oxyhydroxides. These forms of iron are generally insoluble and not readily available to plants (Barman and Das, 2021). The solubility of iron in the soil is influenced by its chemical properties, including its oxidation state, pH, and the presence of other ions (Barman and Das, 2021; Briat and Lebrun, 2021). Iron exists in two oxidation states in soil, ferrous (Fe^{2+}) and ferric (Fe^{3+}). Ferrous iron is soluble and readily available to plants, whereas ferric iron is insoluble and not easily available to plants (Marschner, 2012). Soil pH is a critical factor in the solubility of iron in soil. Iron solubility decreases as soil pH increases, and iron is more available to plants in acidic soils (pH < 7) (Barman and Das, 2021). In alkaline soils, iron tends to form insoluble complexes with hydroxides and carbonates, making it less available to plants. The formation of these complexes is due to the high affinity of ferric iron for hydroxyl and carbonate ions, which leads to the formation of precipitates that are not readily available to plants (Barman and Das, 2021; Briat and Lebrun, 2021). The presence of other ions in the soil can also affect the solubility of iron. For example, the solubility of iron decreases in the presence of phosphate ions due to the formation of insoluble iron phosphate complexes (Marschner, 2012). Similarly, the solubility of iron decreases in the presence of calcium ions due to the formation of insoluble calcium-iron complexes (Kabata-Pendias and Mukherjee, 2017). In addition to chemical factors, physical factors such as soil texture and structure can also affect the solubility of iron in the soil. Iron is more soluble in sandy soils than in clay soils due to the higher mobility of iron in sandy soils. In contrast, iron tends to be fixed more strongly to clay minerals in clay soils, making it less available to plants (Jones et al., 2016).

Fe in Plants and Its Deficiency

Fe plays a crucial role in many processes within plants. It is an essential micronutrient that is required for the production of chlorophyll, which is necessary for photosynthesis. Without adequate Fe, plants cannot produce enough chlorophyll, resulting in yellowing leaves and stunted growth (Rout and Sahoo, 2015). In addition to its role in chlorophyll production, Fe is also involved in other important processes within plants. For example, it is required for the synthesis of heme, which is a component of many enzymes involved in respiration and nitrogen fixation (Bhatla et al., 2018). Fe is also involved in the production of lignin, which is a structural component of cell walls that provides support and protection for plants (Tamaru et al., 2019). Fe is taken up by plant roots in its soluble form, Fe^{2+}, and is transported to the leaves and other parts of the plant via the xylem. However, plants can only absorb a limited amount of Fe from the soil, so it is important to ensure that there is adequate Fe available in the soil. This can be achieved through the addition of Fe-containing fertilizers or by promoting the activity of microorganisms that can solubilize Fe in the soil (Tagliavini and Rombola, 2001). Fe deficiency is a common problem in many agricultural systems, particularly in alkaline soils. Symptoms of Fe deficiency include yellowing leaves, stunted growth, and reduced yields. In severe cases, Fe deficiency can lead to chlorosis and necrosis of plant tissue, which can result in plant death (Fageria et al., 2002).

Microbes-Mediated in Fe Solubilization

Rhizosphere microbes can play an important role in increasing the mobility of Fe in the soil and making it more available for plants. These microbes have the ability to solubilize Fe from insoluble forms, such as Fe oxides and hydroxides, and convert it into a soluble form that can be taken up by plant roots (Colombo et al., 2014; Khoshru et al., 2020). One mechanism by which rhizosphere microbes solubilize Fe is through the production of siderophores. Siderophores are small organic molecules that bind to Fe and make it more soluble (Ansari et al., 2017). Many rhizosphere bacteria, such as *Pseudomonas* and *Bacillus* species, are known to produce siderophores, which can increase the availability of Fe in the soil (Meena et al., 2016). In addition to siderophore production, rhizosphere bacteria can also promote the reduction of Fe to its soluble ferrous form (Fe^{2+}). This is particularly important in aerobic soils, where Fe is more likely to be present in its insoluble ferric form (Fe^{3+}). Some bacteria, such as *Shewanella* and *Geobacter* species, are known to have the ability to reduce Fe^{3+}, making it more available for plant uptake (Zaidi et al., 2021). Rhizosphere bacteria can also promote the dissolution of Fe through the production of organic acids. Organic acids, such as citric acid and malic acid, can chelate Fe and increase its solubility. Many bacterial species, including *Azospirillum* and *Pseudomonas*, are known to produce organic acids that can promote Fe solubility (Daniel et al., 2022). The intervention of rhizosphere bacteria in increasing the mobility and bioavailability of Fe in the soil can have significant benefits for plant growth and productivity. By solubilizing Fe from insoluble forms, promoting its reduction to the ferrous form, and producing organic acids that increase its solubility, these bacteria can improve the availability of this important micronutrient for plants (Rashid at al., 2016). This can help to reduce the occurrence of Fe deficiency in crops, leading to increased yields and improved food security.

Isolation and Screening of FSB

Isolating Fe-solubilizing bacteria (FSB) from soil and rhizosphere can be done using the following techniques. Sampling of soil and rhizosphere of different plants is done from a depth of 0-20 cm. Sampling is done using sterile laboratory tools (including a small laboratory shovel, scissors and plastic bags). In order to obtain rhizospheric FSB, it is necessary to isolate them on media such as TMS-Fe (Tris minimal salt medium, containing insoluble iron) via the serial dilution method (Figure 1).

The modified method by Saravanan et al. (2004) is used to evaluate the solubility of low-soluble iron. The TMS-Fe medium is prepared, which contains 0.1% low-soluble iron sources, including iron phosphate ($FePO_4$), hematite (Fe_2O_3), and goethite (α-FeO(OH)). These iron sources are added separately to the TMS-Fe medium, meaning that only one source is used in each medium, not a mixture of them. The TMS-Fe medium contains 10 g D-glucose, 6.06 g Tris-HCl, 4.68 g NaCl, 1.49 g KCl, 1.07 g NH_4Cl, 0.43 g Na_2SO_4, 0.2 g $MgCl_2 \cdot 2H_2O$, 0.2 g $CaCl_2 \cdot 2H_2O$, and 1 L distilled water. A bacterial overnight suspension grown on NB (Nutrient broth) medium is inoculated by 100 µL (10^8 CFU/mL) to 50 mL of TMS-Fe medium. An autoclaved and un-inoculated medium (containing only 100 µL of sterile NB) is used as a control. The flasks are shaken for 7 days at 28°C and 125 rpm in a shaking incubator. The suspensions are then centrifuged at 6,000 rpm for 10 minutes to separate the soluble iron from bacterial cells, suspended particles, and insoluble iron. The soluble iron is then quantified using AAS (Atomic absorption spectroscopy) methods, and the pH of the liquid supernatant is measured using a pH meter to examine changes in pH (Khoshru et al., 2022).

A similar process for the isolation and screening of zinc-solubilizing bacteria was reported by Khoshru et al. (2023). In general, this process is done first in a solid environment and then in a liquid environment containing the insoluble form of the desired nutrients or elements (such as iron, zinc, phosphorus, potassium, etc.) (Sarikhani et al., 2019; Sarikhani et al., 2018; Sarikhani et al., 2016; Ebrahimi et al., 2019; Nobahar et al., 2017). However, microbial biomass Zn can be considered too, as proposed by other researchers for estimation of K releasing efficiency of bacteria (Ebrahimi et al., 2019). To choose efficient FSB strains, it is necessary to carry out a pot culture or field trial. We expect that there will be a positive correlation between their potentials either in laboratory used media or pot cultures.

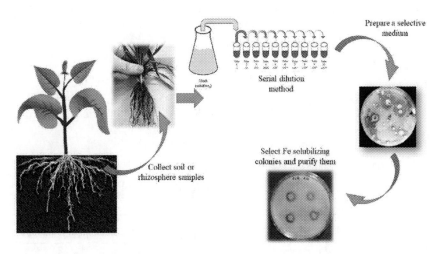

Figure 1. Schematic diagram of the FSB isolation (Khoshru et al., 2022).

Diversity of Fe-Solubilizers

The diversity of Fe-solubilizing microbes (FSMs) in soil is vast and includes many different bacterial and fungal species (Table 1). The ability of these microorganisms to solubilize Fe depends on a range of factors, including soil pH, redox potential, and the availability of organic matter (Boiteau et al., 2019). Bacteria are the most commonly studied group of FSMs in soil. Some of the most well-known genera of FSB include *Pseudomonas*, *Bacillus*, and *Rhizobium*. These bacteria are often found in the rhizosphere of plants, where they can solubilize Fe and make it more available for plant uptake (Compant et al., 2019). Fungi also play an important role in the solubilization of Fe in soil. Many species of fungi, including *Aspergillus*, *Penicillium*, and *Trichoderma*, are known to produce organic acids that chelate Fe and increase its solubility (Vassileva et al., 2022). Other fungal species, such as arbuscular mycorrhizal fungi (AMF), can form symbiotic relationships with plant roots, in which they provide the plant with Fe in exchange for nutrients (Wang et al., 2017). In addition to bacteria and fungi, other groups of microorganisms, such as actinomycetes, algae, and protozoa, have also been shown to solubilize Fe in soil. Actinomycetes are particularly important in acidic soils, where they can solubilize Fe through the production of organic acids (Jog et al., 2016). Algae can also contribute to Fe solubilization in soil, particularly in aquatic environments, where they can produce siderophores and other chelating agents (Subramaniyam et al., 2016). Overall, the diversity of FSMs in soil is vast and includes many different taxonomic groups. The ability of these microorganisms to solubilize Fe varies depending on the specific conditions of the soil environment. Understanding the diversity and function of FSMs in soil is important for developing strategies to improve plant growth and productivity and for promoting sustainable agricultural practices (Ellermann and Arthur, 2017).

Table 1. Biodiversity of microorganisms having Fe solubilization potential

Type of Microbe	Microbe Name	Effect	Reference
Bacteria	*Bacillus subtilis*	Solubilized $FePO_4$	Chen et al., 2019
Bacteria	*Pseudomonas aeruginosa*	Solubilized Fe_2O_3	Liu et al., 2018
Bacteria	*Enterobacter* spp.	Enhanced plant growth and Fe uptake	Han et al., 2020
Bacteria	*Streptomyces* spp.	Solubilized α-FeO(OH)	Ren et al., 2019
Bacteria	*Rhizobium leguminosarum*	Improved Fe availability to plants	Li et al., 2020
Bacteria	*Paenibacillus polymyxa*	Increased plant-growth and Fe uptake	Zhang et al., 2019
Bacteria	*Serratia marcescens*	Solubilized $FePO_4$	Wang et al., 2021
Fungi	*Aspergillus niger*	Solubilized Fe_2O_3	Li et al., 2019
Fungi	*Trichoderma harzianum*	Increased Fe availability to plants	Guo et al., 2018

A Sustainable Approach to Increase the Bioavailability of Iron in Plants   147

Type of Microbe	Microbe Name	Effect	Reference
Fungi	*Penicillium oxalicum*	Solubilized α-FeO(OH)	Lu et al., 2020
Fungi	*Neurospora crassa*	Enhanced plant growth and Fe uptake	Zhang et al., 2021
Fungi	*Cunninghamella elegans*	Solubilized FePO4	Liu et al., 2019
Bacteria-Fungi consortium	*Bacillus subtilis-Trichoderma harzianum*	Improved Fe availability to plants	Gao et al., 2017
Bacteria	*Stenotrophomonas, Pantoea and Agromyces*	Effective in Fe solubilizing from iron phosphate, hematite, and goethite sources	Khoshru et al., (2022

Bacteria as FSB

Fe-solubilizing rhizosphere bacteria are microorganisms that can solubilize Fe from insoluble forms in the soil and make it more available for plant uptake. These bacteria are commonly found in the rhizosphere, which is the region of soil immediately surrounding plant roots, where they interact with plants and play an important role in nutrient cycling (Khoshru et al., 2022). There are many different species of Fe-solubilizing rhizobacteria, including members of the genera *Pseudomonas*, *Bacillus*, *Rhizobium*, and *Azospirillum*, among others (Koul et al., 2019). These bacteria can solubilize Fe through a variety of mechanisms, including the production of organic acids, the secretion of siderophores, and the reduction of Fe^{3+} to Fe^{2+} (Subrahmanyam et al., 2020). Organic acids, such as citric acid and malic acid, can chelate Fe and make it more soluble in the soil. Many FSB are known to produce organic acids, which can increase the availability of Fe for plant uptake (Athukorala, 2021). Siderophores are small molecules that bind to Fe and make it more soluble. Many bacterial species, such as *Pseudomonas* and *Bacillus*, are known to produce siderophores, which can increase the availability of Fe in the soil (Pahari et al., 2017). FSB can also promote the reduction of Fe^{3+} to Fe^{2+}, which is more soluble and available for plant uptake. Some bacteria, such as *Shewanella* and *Geobacter* species, are known to have the ability to reduce Fe in anaerobic soils, making it more available for plant uptake (Estevez-Canales et al., 2015). Khoshru et al., (2022) in a study, investigated the potential of 20 bacteria isolated from the rhizosphere of corn, wheat and sunflower plants for the solubilization of low soluble iron sources. Among the isolates, three promising isolates, ZP15, ZC10, and ZP13, were obtained, which were effective in solubilizing iron phosphate, hematite, and goethite sources. They produced various organic acids and siderophores for iron solubilization. Molecular and biochemical analysis of the isolates showed that they belonged to *Stenotrophomonas*, *Pantoea* and *Agromyces* genera.

Fungi as FSF

Fungi are also important microorganisms that can solubilize Fe in soil. Many species of fungi, including *Aspergillus*, *Penicillium*, and *Trichoderma*, are known to produce organic acids that can chelate Fe and make it more soluble (Vassileva et al., 2022). Other fungal species, such as arbuscular mycorrhizal fungi, can form symbiotic relationships with plant roots, in which they provide the plant with Fe in exchange for nutrients. AMF are a particularly important group of fungi that can solubilize Fe in soil. These fungi form symbiotic associations with plant roots, in which they colonize the root system and provide the plant with nutrients, including Fe (Trouvelot et al., 2015). AMF can solubilize Fe through the production of organic acids, such as citric acid, which can chelate Fe and increase its solubility (Andrino et al., 2021). AMF can also produce siderophores, which can bind to Fe and make it more available for plant uptake (Mishra et al., 2016). In addition to AMF, other fungal species are also known to solubilize Fe in the soil. For example, some species of *Penicillium* and *Aspergillus* are able to solubilize Fe from insoluble forms, such as Fe oxides and hydroxides. These fungi can produce organic acids, such as gluconic acid and citric acid, which can chelate Fe and increase its solubility (Andrino et al., 2021).

Mechanisms of Fe Solubilization

Microorganisms can solubilize Fe in soil through a variety of mechanisms (Table 2). Here are some of the most common mechanisms used by microbes to solubilize Fe:

1. Production of organic acids: One of the most common mechanisms used by microbes to solubilize Fe is through the production of organic acids. Organic acids, such as citric acid and malic acid, can chelate Fe and make it more soluble in the soil. Many bacterial and fungal species are known to produce organic acids that can increase the

availability of Fe in the soil (Sarikhani et al., 2020; Andrino et al., 2021).

2. Secretion of siderophores: Siderophores are small molecules that bind to Fe and make it more soluble. Many bacterial and fungal species are known to secrete siderophores, which can increase the availability of Fe in the soil (Saha et al., 2016; Roskova et al., 2022).

3. Reduction of Fe^{3+}: Some bacteria, such as *Shewanella* and *Geobacter* species, are known to have the ability to reduce Fe^{3+} in anaerobic soils, making it more available for plant uptake. This reduction can be an important mechanism for solubilizing Fe in soil (Cain and Smith, 2021; Li et al., 2022).

4. Acidification of the soil: Microorganisms can also solubilize Fe by acidifying the soil. This acidification can increase the solubility of Fe by dissolving Fe oxides and hydroxides. Many bacterial and fungal species are known to produce acids that can decrease the pH of the soil and increase the availability of Fe (Fujii et al., 2018; Khoshru et al., 2022).

5. Oxidation of Fe: Some bacteria and fungi have the ability to oxidize Fe^{2+} to Fe^{3+}, which can then be solubilized by other microorganisms. In this case Iron is not available for microorganisms. This process is particularly important in aerobic soils, where Fe^{2+} is more soluble than Fe^{3+} (Bonneville et al., 2016).

Table 2. Mechanisms of iron solubilization
by different microorganisms in soil

Microorganism type	Microbe Name	Solubilization mechanism	Reference
Bacteria	*Pseudomonas fluorescens*	Production of siderophores	Khan et al., 2014
	Bacillus megaterium	Production of organic acids	Nautiyal et al., 2010
	Bacillus subtilis	Reduction of Fe(III) to Fe(II)	Nayak et al., 2016
	Streptomyces griseus	Production of organic acids	Liu et al., 2013
	Rhizobium leguminosarum	Production of siderophores	Gupta et al., 2012
Fungi	*Aspergillus niger*	Production of organic acids	Gadd and Pan, 2016
	Trichoderma viride	Production of organic acids and siderophores	Datta et al., 2015

Table 2. (Continued)

Microorganism type	Microbe Name	Solubilization mechanism	Reference
	Penicillium citrinum	Production of organic acids and siderophores	Kumar et al., 2014
	Fusarium oxysporum	Production of organic acids and siderophores	Singh et al., 2016
	Neurospora crassa	Production of organic acids	Liu et al., 2014
	Phanerochaete chrysosporium	Production of organic acids	Yuan et al., 2011

Microbes use diverse mechanisms to solubilize Fe in the soil, which can vary based on the specific microorganisms and environmental conditions. By comprehending these mechanisms, it may be feasible to devise strategies to enhance the accessibility of Fe in soil and encourage sustainable agricultural practices (de Souza et al., 2015).

Role of Fe-Solubilizing in Sustainable Agriculture

FSMs play an important role in sustainable agriculture by improving soil health, promoting plant growth, and reducing the need for synthetic fertilizers. Here are some ways in which FSMs can contribute to sustainable agriculture: 1- Improving soil health: FEMs can help to improve soil health by promoting the solubilization of Fe and other nutrients. By increasing the availability of these nutrients, these microbes can improve soil fertility and reduce the need for synthetic fertilizers (Kumar and Verma, 2019; Devi et al., 2020). 2- Enhancing plant growth: FEMs can also enhance plant growth by making Fe more available for plant uptake. Fe is an essential nutrient for plant growth, and deficiencies can lead to stunted growth and reduced yields. By improving the availability of Fe, these microbes can promote plant growth and increase yields (Kartik et al., 2022). 3- Reducing the need for synthetic fertilizers: FEMs can reduce the need for synthetic fertilizers, which can be costly and have negative environmental impacts. By promoting the solubilization of Fe and other nutrients, these microbes can reduce the need for synthetic fertilizers and promote more sustainable agricultural practices (Wang et al., 2020). 4- Promoting soil sustainability: FSMs can also help to promote soil sustainability by improving soil structure, increasing water retention, and reducing erosion. By improving soil health and fertility, these microbes can

A Sustainable Approach to Increase the Bioavailability of Iron in Plants   151

help to maintain the long-term productivity and sustainability of agricultural soils (Kumar et al., 2015).

Generally, the role of FEMs in sustainable agriculture is significant. These microbes can help to improve soil health, promote plant growth, reduce the need for synthetic fertilizers, and promote soil sustainability. By understanding the mechanisms by which these microbes solubilize Fe, it may be possible to develop strategies to enhance their activity and promote their use in agricultural systems. For example, inoculating soils with FEMs or using biofertilizers that contain these microbes can be effective ways to enhance their activity and promote sustainable agriculture (Reddy et al., 2020).

Conclusion and Future Perspectives

In conclusion, FSMs, including FSB and FSF, play an important role in solubilizing Fe in soil and making it more available to plants. These microbes use a variety of mechanisms, including the production of organic acids, the secretion of siderophores, and the reduction of Fe, among others. By promoting the solubilization of Fe and other nutrients, these microbes can improve soil health, enhance plant growth, reduce the need for synthetic fertilizers, and promote soil sustainability. In the future, there is a need for further research to better understand the diversity and function of FSMs in soil. This research can help to identify new microbial strains that can be used to improve plant growth and productivity, and to promote sustainable agricultural practices. In addition, there is a need to develop effective strategies for enhancing the activity of FSMs in soil. This may include the use of biofertilizers or composts containing FSMs, or the development of microbial inoculants that can be applied directly to soils. Overall, the use of FSMs in agriculture has the potential to improve soil health, promote sustainable agriculture, and enhance food security. By promoting the availability of Fe and other nutrients, these microbes can help to increase crop yields, reduce the need for synthetic fertilizers, and promote more sustainable agricultural practices.

References

Andrino A, Guggenberger G, Kernchen S, Mikutta R, Sauheitl L, Boy J. Production of organic acids by arbuscular mycorrhizal fungi and their contribution in the mobilization of phosphorus bound to iron oxides. *Front Plant Sci.* (2021)12:661842.

Ansari RA, Mahmood I, Rizvi R, Sumbul A, Safiuddin. Siderophores: augmentation of soil health and crop productivity. In: Kumar V, Kumar M, Sharma S, Prasad R, editors. *Probiotics in Agroecosystem.* Springer; (2017):291-312.

Arora NK, Fatima T, Mishra I, Verma M, Mishra J, Mishra V. Environmental sustainability: challenges and viable solutions. *Environ Sustain.* (2018)1:309-340.

Athukorala ASN. Solubilization of micronutrients using indigenous microorganisms. In: Giri B, Varma A, Prasad R, editors. *Microbial Technology for Sustainable Environment.* Springer; (2021):365-417.

Barman D, Das A. Iron in soil: forms, functions, and availability in the environment. In: Barman D, Das A, editors. *Iron in Plants and Soil Microorganisms.* Springer; (2021:1-32.

Bhatla SC, Lal AM,Kathpalia R, Bhatla SC. Plant mineral nutrition. In: Tripathi SB, editor. *Plant Physiology, Development and Metabolism.* Springer; (2018):37-81.

Boiteau RM, Fansler SJ, Farris Y, Shaw JB, Koppenaal DW, Pasa-Tolic L, Jansson JK. Siderophore profiling of co-habitating soil bacteria by ultra-high resolution mass spectrometry. *Metallomics.* (2019)11(1):166-175.

Bonneville S, Bray AW, Benning LG. Structural Fe (II) oxidation in biotite by an ectomycorrhizal fungi drives mechanical forcing. *Environ Sci Technol.* (2016);50(11):5589-5596.

Briat JF, Lebrun M. Iron: physiological functions and metabolism in plants. In: Lennarz WJ, Lane MD, editors. Encyclopedia of Biological Chemistry. *Academic Press*; (2021:316-322.

Cain TJ, Smith AT. Ferric iron reductases and their contribution to unicellular ferrous iron uptake. *J Inorg Biochem.* (2021);218:111407.

Chen X, Chen Y, Zhang L, Lin Q. Bacillus subtilis as a potential iron-solubilizing bacterium: Solubilization mechanisms and colonization ability to improve maize growth under calcareous soil conditions. *J Appl Microbiol.* (2019);127(6):1644-1656.

Chen X, Li X, Zhang Y. Enterobacter sp. YSU-8 solubilizes insoluble iron minerals and promotes plant growth in calcareous soil. *J Basic Microbiol.* (2020);60(5):385-396.

Chen Y, Barak P. Iron nutrition of plants in calcareous soils. *Adv Agron.* (1982);35:217-240.

Chen Y, Liao H. Iron-regulated transcription factors and their roles in signaling and plant growth response. *Plant Cell Physiol.* (2016);57(1):8-16.

Chopra A, Vandana UK, Rahi P, Satpute S, Mazumder PB. Plant growth promoting potential of Brevibacterium sediminis A6 isolated from the tea rhizosphere of Assam, India. Biocatal *Agric Biotechnol.* (2020);27:101610.

Colombo C, Palumbo G, He JZ, Pinton R, Cesco S. Review on iron availability in soil: interaction of Fe minerals, plants, and microbes. *J Soils Sediments.* (2014);14:538-548.

A Sustainable Approach to Increase the Bioavailability of Iron in Plants   153

Compant S, Samad A, Faist H, Sessitsch A. A review on the plant microbiome: ecology,functions, and emerging trends in microbial application. *J Adv Res*. (2019);19:29-37.

Crowley DE, Wang YC, Reid CPP, Szaniszlo PJ. Mechanisms of iron acquisition from siderophores by microorganisms and plants. In: Barton LL, Hemming BC, editors. *Iron Nutrition in Plants and Rhizospheric Microorganisms.* Springer Netherlands; (1991):213-232.

Daniel AI, Fadaka AO, Gokul A, Bakare OO, Aina O, Fisher S, Klein A. Biofertilizer: The Future of Food Security and Food Safety. *Microorganisms*. (2022);10(5):1220.

Datta R, Baraniya D, Chandra P, Kim K. Agricultural wastes and their potential utilization in bioremediation of heavy metals-contaminated soils. *Environ Sci Pollut Res Int*. (2015);22(19):14362-14381.

De Souza R, Meyer J, Schoenfeld R, da Costa PB, Passaglia LM. Characterization of plant growth-promoting bacteria associated with rice cropped in iron-stressed soils. *Ann Microbiol*. (2015);65:951-964.

Devi R, Kaur T, Kour D, Rana KL, Yadav A, Yadav AN. Beneficial fungal communities from different habitats and their roles in plant growth promotion and soil health. *MicrobialBiosyst*. (2020);5(1):21-47.

Ebrahimi M, Safari Sinegani AA, Sarikhani MR, Aliasgharzad N. Assessment of Soluble and Biomass K in Culture Medium Is a Reliable Tool for Estimation of K Releasing Efficiency of Bacteria. *Geomicrobiol J*. (2019);36(10):873-880.

Ellermann M, Arthur JC. Siderophore-mediated iron acquisition and modulation of host-bacterial interactions. *Free Radic Biol Med*. (2017);105:68-78.

Estevez-Canales M, Kuzume A, Borjas Z, Füeg M, Lovley D, Wandlowski T, Esteve-Núñez A. A severe reduction in the cytochrome C content of Geobacter sulfurreducens eliminates its capacity for extracellular electron transfer. *Environ Microbiol Rep*. (2015);7(2):219-226.

Fageria NK, Baligar VC, Clark RB. Micronutrients in crop production. *Adv Agron*. (2002);77:185-268.

Fujii K, Shibata M, Kitajima K, Ichie T, Kitayama K, Turner BL. Plant–soil interactions maintain biodiversity and functions of tropical forest ecosystems. *Ecol Res*. (2018);33:149-160.

Gadd GM, Pan X. Bioremedial potential of microbial mechanisms of metal mobilization and immobilization. *Curr Opin Biotechnol*. (2016);38:10-17.

Gambrell RP. Trace and toxic metals in wetlands—a review. *J Environ Qual*. (1994);23(5):883-891.

Gao X, Lu L, Li X. Bacillus subtilis-Trichoderma harzianum consortium improves maize root system development and nutrient uptake. *J Appl Microbiol*. (2017);122(5):1142-1153.

Graham RD. Micronutrient deficiencies in crops and their global significance. Springer Netherlands; (2008): 41-61.

Gregory PJ, Wahbi A, Adu-Gyamfi J, Heiling M, Gruber R, Joy EJ, Broadley MR. Approaches to reduce zinc and iron deficits in food systems. *Global Food Security*. (2017);15:1-10.

Guo Y, Li X, Zhang Y. Trichoderma harzianum enhances iron acquisition efficiency of maize plants. *J Basic Microbiol*. (2018);58(4):303-312.

Gupta R, Singh R, Pandey R. Microbial degradation of xenobiotic compounds. *Environ Sci Pollut Res Int*. (2012);19(9):3793-3802.

Han Y, Li X, Yang Y. Enterobacter sp. YSU-5 improves iron nutrition and plant growth in calcareous soil. *Soil Sci Plant Nutr*. (2020);66(2):280-287.

Hayat R, Ali S, Amara U, Khalid R, Ahmed I. Soil beneficial bacteria and their role in plant growth promotion: a review. *Ann Microbiol*. (2010);60:579-598.

Jog R, Nareshkumar G, Rajkumar S. Enhancing soil health and plant growth promotion by actinomycetes. In: Singh D, Singh H, Prabha R, editors. Plant growth promoting actinobacteria: A new avenue for enhancing the productivity and soil fertility of grain legumes. Springer; (2016): 33-45.

Kabata-Pendias A, Mukherjee AB. *Trace elements from soil to human.* Springer; (2017).

Kartik VP, Chandwani S, Amaresan N. Augmenting the bioavailability of iron in rice grains from field soils through the application of iron-solubilizing bacteria. *Lett Appl Microbiol*. (2022);69(2):185-192.

Khan MS, Zaidi A, Ahemad M. Ochrobactrum ciceri sp. nov., an endophytic bacterium isolated from root nodules of Cicer arietinum. *Int J Syst Evol Microbiol*. (2013);63(Pt 5):1873-1878.

Khoshru B, Mitra D, Joshi K, Adhikari P, Rion MSI, Fadiji AE, Keswani C. Decrypting the multi-functional biological activators and inducers of defense responses against biotic stresses in plants. *Heliyon*. (2023);9(1):e13825.

Khoshru B, Mitra D, Khoshmanzar E, Myo EM, Uniyal N, Mahakur B, Rani A. Current scenario and future prospects of plant growth-promoting rhizobacteria: An economic valuable resource for the agriculture revival under stressful conditions. *J Plant Nutr*. (2020);43(20):3062-3092.

Khoshru B, Moharramnejad S, Gharajeh NH, Asgari Lajayer B, Ghorbanpour M. Plant microbiome and its important in stressful agriculture. In: Maheshwari D, Saraf M, Aeron A, editors. *Plant microbiome paradigm.* Springer; (2020): 13-48.

Khoshru B, Sarikhani MR, Reyhanitabar A, Oustan S, Malboobi MA. Evaluation of the Ability of Rhizobacterial Isolates to Solubilize Sparingly Soluble Iron Under In-vitro Conditions. *Geomicrobiol J*. (2022);39(8):804-815.

Khoshru B, Sarikhani MR, Reyhanitabar A, Oustan S, Malboobi MA. Evaluation of the potential of rhizobacteria in supplying nutrients of Zea mays L. plant with a focus on zinc. *J Soil Sci Plant Nutr*.(2023);1-14.

Koul B, Singh S, Dhanjal DS, Singh J. Plant growth-promoting rhizobacteria (PGPRs): a fruitful resource. In: Singh D, Singh H, Prabha R, editors. *Microbial Interventions in Agriculture and Environment: Volume 3: Soil and Crop Health Management.* Springer; (2019): 83-127.

Kumar A, Singh R, Yadav A. Bioremediation of heavy metals using microbes. *Adv Bioresearch*. (2014);5(2):57-64.

Kumar A, Verma JP. The role of microbes to improve crop productivity and soil health. In: Singh D, Singh H, Prabha R, editors. *Ecological Wisdom Inspired Restoration Engineering.* Springer; (2019): 249-265.

A Sustainable Approach to Increase the Bioavailability of Iron in Plants 155

Kumar A, Bahadur I, Maurya BR, Raghuwanshi R, Meena VS, Singh DK, Dixit J. Does a plant growth-promoting rhizobacteria enhance agricultural sustainability. *J Pure Appl Microbiol*. (2015);9(1):715-724.

Lemanceau P, Expert D, Gaymard F, Bakker PAHM, Briat JF. Role of iron in plant–microbe interactions. *Adv Bot Res*. (2009);51:491-549.

Li J, Li X, Wang Y. Rhizobium leguminosarum promotes iron acquisition and increases the yield and quality of peanut (Arachis hypogaea L.). *Ann Appl Biol*. (2020); 177(1):105-118.

Li L, Li X, Li Y. Aspergillus niger R21 enhances iron availability in calcareous soil and promotes maize growth. *J Basic Microbiol*. (2019);59(11):1092-1102.

Li H, Ding S, Song W, Zhang Y, Ding J, Lu J. Iron reduction characteristics and kinetic analysis of Comamonas testosteroni Y1: a potential iron-reduction bacteria. *Biochem Eng J*. (2022);177:108256.

Lindsay WL, Schwab AP. The chemistry of iron in soils and its availability to plants. *J Plant Nutr*. (1982);5(4-7):821-840.

Liu H, Li M, Zhao Y, Zhang R, Du G, Chen J, Zhou J. Effects of soil type and soil management on the diversity and abundance of soil fungi. *J Agric Sci*. (2014); 6(8):126-135.

Liu J, Wang H, Wu Z. Solubilization of insoluble iron minerals by Pseudomonas aeruginosa and its plant growth-promoting activities. *Pedosphere*. (2018);28(6):941-950.

Liu Y, Li X, Zhang Y. Cunninghamella elegans enhances iron availability and promotes plant growth in calcareous soil. *Plant Soil*. (2019);435(1-2):91-105.

Lu L, Li X, Zhang Y. Penicillium oxalicum XM311 enhances iron availability and promotes plant growth in calcareous soil. *J Basic Microbiol*. (2020);60(1):60-72.

Mabrouk Y, Hemissi I, Salem IB, Mejri S, Saidi M, Belhadj O. Potential of rhizobia in improving nitrogen fixation and yields of legumes. *Symbiosis*. (2018);107(73495):1-6.

Marschner H. Marschner's mineral nutrition of higher plants. *Academic Press*; (2012).

Meena VS, Maurya BR, Meena SK, Meena RK, Kumar A, Verma JP, Singh NP. Can Bacillus species enhance nutrient availability in agricultural soils? *Bacilli and agrobiotechnology*. (2016):367-395.

Mishra V, Gupta A, Kaur P, Singh S, Singh N, Gehlot P, Singh J. Synergistic effects of Arbuscular mycorrhizal fungi and plant growth promoting rhizobacteria in bioremediation of iron contaminated soils. *International journal of phytoremediation* (2016);18(7):697-703.

Murakami C, Tanaka AR, Sato Y, Kimura Y, Morimoto K. Easy detection of siderophore production in diluted growth media using an improved CAS reagent. *Journal of Microbiological Methods*. (2021);189:106310.

Mwangi MN, Mzembe G, Moya E, Verhoef H. Iron deficiency anaemia in sub-Saharan Africa: a review of current evidence and primary care recommendations for high-risk groups. *The Lancet Haematology*. (2021);8(10):e732-e743.

Nair KP, Nair KP. Soil fertility and nutrient management. In Intelligent Soil Management for Sustainable Agriculture: *The Nutrient Buffer Power Concept* (2019): 165-189.

Nautiyal CS, Bhadauria S, Kumar P, Lal H, Mondal R, Verma D. Stress induced phosphate solubilization in bacteria isolated from alkaline soils. *FEMS Microbiol Lett.* (2010);269(1):98-107.

Nayak S, Pradhan S, Das S. Iron solubilization from magnetite in the presence of Bacillus subtilis. *J Environ Chem Eng.* (2016);4(1):332-340.

Nobahar A, Sarikhani MR, Chalabianlou N. Buffering capacity affects phosphorous solubilization assays in rhizobacteria. *Rhizosphere.* (2017);4:119-125.

Oliver MA, Gregory PJ. Soil, food security and human health: a review. *Eur J Soil Sci.* (2015);66(2):257-276.

Pahari A, Pradhan A, Nayak SK, Mishra BB. Bacterial siderophore as a plant growth promoter. In: Microbial Biotechnology: Volume 1. *Applications in Agriculture and Environment.* (2017):163-180.

Pearsall WH, Mortimer CH. Oxidation-reduction potentials in waterlogged soils, natural waters and muds. *J Ecol.* (1939);27(3):483-501.

Rashid MI, Mujawar LH, Shahzad T, Almeelbi T, Ismail IM, Oves M. Bacteria and fungi can contribute to nutrients bioavailability and aggregate formation in degraded soils. *Microbiol Res.* (2016);183:26-41.

Reddy GC, Goyal RK, Puranik S, Waghmar V, Vikram KV, Sruthy KS. Biofertilizers toward sustainable agricultural development. In: *Plant microbe symbiosis.* Springer; (2020):115-128.

Ren L, Li Y, Yang X. Solubilization of iron oxide (α-FeO(OH)) by Streptomyces sp. 7-145 and its plant growth-promoting effects. *J Basic Microbiol.* (2019);59(4):371-380.

Roskova Z, Skarohlid R, McGachy L. Siderophores: an alternative bioremediation strategy?. *Sci Total Environ.* (2022);153144.

Rout GR, Sahoo S. Role of iron in plant growth and metabolism. *Rev Agric Sci.* (2015);3:1-24.

Saha M, Sarkar S, Sarkar B, Sharma BK, Bhattacharjee S, Tribedi P. Microbial siderophores and their potential applications: a review. *Environ Sci Pollut Res Int.* (2016);23(5):3984-3999.

Sarikhani MR, Aliasgharzad N, Khoshru B. P solubilizing potential of some plant growth promoting bacteria used as ingredient in phosphatic biofertilizers with emphasis on growth promotion of Zea mays L. *Geomicrobiol J.* (2020);37(4):343-352.

Sarikhani MR, Khoshru B, Greiner R. Isolation and identification of temperature tolerant phosphate solubilizing bacteria as a potential microbial fertilizer. *World J Microbiol Biotechnol.* (2019);35(8):126.

Sarikhani MR, Khoshru B, Oustan Sh. Efficiency of some bacterial strains on potassium release from micas and phosphate solubilization under in vitro conditions. *Geomicrobiol J.* (2016);33(9):832-838.

Sarikhani MR, Oustan S, Ebrahimi M, Aliasgharzad N. Isolation and identification of potassium-releasing bacteria in soil and assessment of their ability to release potassium for plants. *Eur J Soil Sci.* (2018);69(6):1078-1086.

Singh A, Singh R, Singh PK. Microbial production of indole acetic acid and solubilization of inorganic phosphates by endophytic fungi isolated from medicinal plants. *J Microbiol Biotechnol Res.* (2016);6(4):22-29.

A Sustainable Approach to Increase the Bioavailability of Iron in Plants   157

Smith MR, Golden CD, Myers SS. Potential rise in iron deficiency due to future anthropogenic carbon dioxide emissions. *GeoHealth*. (2017);1(6):248-257.

Strawn DG, Bohn HL, O'Connor GA. Soil chemistry. *John Wiley & Sons*; (2020).

Subrahmanyam G, Kumar A, Sandilya SP, Chutia M, Yadav AN. Diversity, plant growth promoting attributes, and agricultural applications of rhizospheric microbes. In: *Plant microbiomes for sustainable agriculture.* Springer; (2020):1-52.

Subramaniyam V, Subashchandrabose SR, Thavamani P, Chen Z, Krishnamurti GSR, Naidu R, Megharaj M. Toxicity and bioaccumulation of iron in soil microalgae. *J Appl Phycol*. (2016);28:2767-2776.

Tagliavini M, Rombola AD. Iron deficiency and chlorosis in orchard and vineyard ecosystems. *Eur J Agron*. (2001);15(2):71-92.

Tamaru Y, Yoshida M, Eltis LD, Goodell B. Multiple iron reduction by methoxylated phenolic lignin structures and the generation of reactive oxygen species by lignocellulose surfaces. *Int J Biol Macromol*. (2019);128:340-346.

Thapa S, Bhandari A, Ghimire R, Xue Q, Kidwaro F, Ghatrehsamani S,...& Goodwin M. Managing micronutrients for improving soil fertility, health, and soybean yield. *Sustainability*. (2021);13(21):11766.

Trouvelot S, Bonneau L, Redecker D, Van Tuinen D, Adrian M, Wipf D. Arbuscular mycorrhiza symbiosis in viticulture: a review. *Agron Sustain Dev*. (2015);35:1449-1467.

Vassileva M, Mendes GDO, Deriu MA, Benedetto GD, Flor-Peregrin E, Mocali S, et al. Fungi, P-solubilization, and plant nutrition. *Microorganisms*. (2022);10(9):1716.

Voroney RP, Winter J. Soil organic matter: a gift from nature for agricultural productivity. In: *Soil Health and Intensification of Agroecosystems.* Springer; (2019): 29-44.

Vose PB. Iron nutrition in plants: a world overview. *Journal of plant nutrition*. (1982);5(4-7):233-249.

Wang Y, Li X, Zhang Y. Pseudomonas aeruginosa-Aspergillus niger consortium enhances iron uptake and plant growth in calcareous soil. *J Basic Microbiol*. (2018);58(12):1016-1026.

Wang Y, Zhang Y, Li X. Serratia marcescens SYBC-1, a potential iron-solubilizing bacterium for improving plant growth and iron uptake. *J Basic Microbiol*. (2021);61(4):313-324.

Wang J, Li R, Zhang H, Wei G, Li Z. Beneficial bacteria activate nutrients and promote wheat growth under conditions of reduced fertilizer application. *BMC microbiology*. (2020);20:1-12.

Wang W, Shi J, Xie Q, Jiang Y, Yu N, Wang E. Nutrient exchange and regulation in arbuscular mycorrhizal symbiosis. *Molecular plant*. (2017);10(9):1147-1158.

Chapter 7

Azotobacter as a Biofertilizer in Sustainable Agriculture

Mohammad Reza Sarikhani[*], PhD and Mitra Ebrahimi, PhD

Department of Soil Science, Faculty of Agriculture, University of Tabriz, Tabriz, Iran

Abstract

Current soil management strategies mainly depend on inorganic chemical fertilizers, which pose a serious threat to human health and the environment. Therefore, it is necessary to increase agricultural productivity to meet the food needs of the growing population. The human population of the world continues to increase, creating a significant challenge in ensuring food security, as soil nutrients and fertility become limited and decrease over time. Application of rhizospheric bacteria is one of the most promising strategies to achieve a sustainable agriculture system. Plant-associated microbes with plant growth-promoting properties have great potential to solve these challenges and play an important role in increasing plant biomass and crop yield. Eco-friendly approaches induce a wide range of applications of beneficial microbes leading to improved nutrient uptake, plant growth, and plant tolerance to biotic and abiotic stresses. Among the plant growth promoting rhizobacteria, *Azotobacter* genus is considered to improve plant health. It is a free–living N_2– fixer diazotroph that has several beneficial effects on the crop growth and yield. Various mechanisms such as nitrogen fixation, plant hormone production, phosphate dissolution, etc., play a role in improving plant health in *Azotobacter* species inoculated plants. Although many studies have been conducted

---

[*] Corresponding Author's Email: rsarikhani@yahoo.com.

In: Biofertilizers
Editor: Philip L. Bevis
ISBN: 979-8-89113-082-1
© 2023 Nova Science Publishers, Inc.

to isolate and identify diazototrophs in sustainable agriculture, the mechanisms of action and prospects of these rhizobacteria in sustainable agriculture remain largely unknown. This study mainly focuses on the Azotobacters and assesses their diversity, action mechanisms, ecological significances and its biotechnological applications. Such data is valuable in deciding their potential and assessing their prospects in promoting sustainable agricultural systems. The use of *Azotobacter* may be a promising method to create and meet the N requirement of the growing crop without causing any environmental risk. This review critically examines the current status of using *Azotobacter* strains as biofertilizers and the important role played by these beneficial microbes in maintaining soil fertility and increasing crop productivity.

Keywords: Azotobacteria, biofertilizer, BNF, nitrogen (N), PGPR, sustainable agriculture

Introduction

Nitrogen (N) is required by all living creatures and is found in all soils. This element is the seventh most abundant element in the world and is the most abundant element in the Earth's atmosphere, making up about 78% (4000 trillion tons) of our atmosphere. N is the most required nutrient in plants. It is a key component of vital organic molecules such as nucleic acids, amino acids and proteins. In short, nitrogen is found in every ecosystem and in every part of the global environment (Anas et al., 2020).

In most agro-ecosystems, N is often the limiting nutrient that dictates crop production. In spite of its presence in large amounts in the form of N_2, plants cannot utilize atmospheric nitrogen (N_2). Application of chemical fertilizer is one approach to meet plant N demand. Intensive agriculture depend on the significant application of nitrogen fertilizers, along with other essential nutrients to maximize crop productivity. Overall, the use of synthetic N-based fertilizers is estimated to produce almost half of the global food supply, and the use of N fertilizers is projected to increase from 80 to 180 million tons by 2050 (Bindraban et al., 2015; Benghzial et al., 2023).

It is estimated that up to fifty percent of conventional N-based fertilizer applications are subject to soil and environmental loss (Anas et al., 2020). This can significantly lead to economic and environmental issues such as increasing greenhouse gas emissions, reducing non-renewable resources, and leaching nitrates into groundwater and surface water, which can cause destructive effects such as water eutrophication. Therefore, there is a need for continued use of nitrogen fertilizers to meet the challenges of agricultural sustainability, which include better crop nutrition and productivity required for an ever-increasing world population. Above all, the safe provision of soil ecosystem services is undoubtedly essential to ensure the sustainability of agro-ecosystems (Benghzial et al., 2023).

Nevertheless, providing sufficient nitrogen to the plant remains a challenge in the case of such a highly mobile nutrient in the soil. Atmospheric nitrogen (N_2) can be fixed chemically (by Haber-Bosch process) or biologically (by diazotrophs) to ammonia. Chemical nitrogen fixation (CNF) needs high amount of energy and can be performed in high temperature and high pressure while biological nitrogen fixation (BNF) is being done in diazotroph cells and in normal condition using nitrogenase enzymes.

In this regard, biologically fixed nitrogen has been the main input of nitrogen in agricultural ecosystems. Nitrogen-fixing bacteria (NFB) live both in plant tissues (such as nodules, roots) and at the soil-root rhizosphere interface, and thus can provide significant amounts of N for plant growth. This is principally due to the microbial process of "biological nitrogen fixation" through bacterial enzymatic conversion of atmospheric N_2 to ammonia (NH_3). BNF can be an environmentally acceptable supplement or alternative to mineral N fertilizers (Kumar et al., 2022). Almost wide range of bacteria known as diazotophs, one of them is *Azotobacter*. Azotobacteria are free-living or non-symbiotic bacteria, heterotrophic, aerobic and gram-negative. These bacteria are able to fix nitrogen and also produce plant growth stimulants such as hormones, vitamins, etc. Azotobacteria have been used as a biofertilizer for more than a century. This genus was first identified in 1901 by Martinus Beijerinck. *Azotobacter* belongs to the family Pseudomonadaceae/Azotobacteraceae and class Gammaproteobacteria, which is common in soils sampled from across the world (Das, 2019; Sumbul et al., 2020).

N in Soils and Its Availability

All living cells need nutrients for their growth. Seventeen nutrients are essential for proper plant growth. Each plant nutrient is equally important to the plant, but each is needed in very different amounts. These 17 nutrients based on plant requirement are divided into two groups (macronutrients and micronutrients), and in some cases they are divided into three groups (including major, minor and micronutrients). Nutrients such as (N:nitrogen, P:phosphorus and K:potassium) are considered as major nutrients; while (Ca:calcium, Mg:magnesium and S:sulphur) are considered as minor nutrients, and the rest including (B:boron, Cl:chlorine, Cu:copper, Fe:iron, Mn:manganese, Mo:molybdenum, Zn:zinc and Ni:nickel) known as micronutrients. Besides the 14 nutrients listed above, plants require C:carbon, H:hydrogen and O:oxygen, which come from air and water (Marschner et al., 2023; Brahmaprakash and Sahu, 2012).

Most plants take up nutrients as inorganic ions regardless of what form of that are added to the soil. Nitrogen is one of the essential components of life, required to make proteins and DNA, and despite its abundant presence in the atmosphere, only limited reserves of inorganic soil nitrogen are available to plants, mainly in the form of nitrate (NO_3^-) and ammonium (NH_4^+). Therefore, agricultural yield is often limited by nitrogen availability (Pankievicz et al., 2019; Marschner et al., 2023). In the soil system, N exists in many forms, and changes or transforms very easily from one form to another. A simplified N cycle is shown in Figure 1, which displays the route N follows in and out of the soil system. The N cycle is influenced by biological processes. Biological processes, sequentially, are influenced by prevailing climatic conditions along with physical and chemical properties of soil. Organic N compounds, and inorganic form of N (NH_4^+ and NO_3^-) are major forms of N in the soil, and the inorganic form of soil N is only a small fraction of the total soil N. Maximum of the N in a surface soil is present as organic N in different N fractions (e.g., proteins, amino sugars, purine and pyrimidime derivates, etc.), and these fractions are susceptible to various transformation processes (Paul, 2007).

N is extensively spread throughout the lithosphere, atmosphere, hydrosphere, and biosphere. Unlike the other two major plant nutrients, phosphorus (P) and potassium (K), rock deposits of N do not exist in the lithosphere, and therefore N fertilizer is made from the conversion of N_2 into reactive forms (NH_3). It is remarkable that only a very small portion of this N is present in the soil (approximately the first meter of the earth's crust), mostly as organic forms. The total N content of surface mineral soils normally ranges between 0.05 and 0.2 percent, corresponding to approximately 1750 to 7000 kg N/ha in the plough layer. Of this total N content only a small proportion, in most cases less than five percent, is directly available to plants, mainly as nitrate and ammonium. Organic N, being the rest, gradually becomes available through mineralization (Hofman and Van Cleemput, 2004).

In general, N is easily transformed from one state to another state of reduction or oxidation and is readily distributed by hydrologic and atmospheric processes. In agro-ecosystems, N is present in many forms covering a range of valence states from −3 to +5 in ammonia and nitrate, respectively (Hofman and Van Cleemput, 2004; Marschner et al., 2023). Deficiency of nitrogen is often a major limiting factor for crop production. N is an essential plant nutrient that is widely used as N-fertilizer to improve the yield of important agricultural crops. A remarkable substitute to avoid or reduce the use of N-fertilizers could be the exploitation of plant growth promoting bacteria (PGPB), which are able to increase the growth and yield of many plant species, some of which are of agronomic and ecological importance.

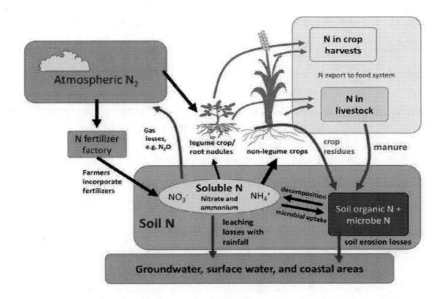

Figure 1. Schematic of Nitrogen cycle and its form (adapted from https://serc.carleton.edu/integrate/teaching_materials/food_supply/student_materials/1175).

N in Plants and Its Deficiency

Nitrogen, as an essential nutrients and component of all living cells, is required to make amino acids, proteins, nucleic acids, and chlorophyll, and even though being abundant in the atmosphere, only limited reserves of inorganic soil nitrogen (NO_3^- and NH_4^+) are available to plants (Marschner et al., 2023).

Consequently, crop yields in agriculture are often limited by nitrogen availability (Pankievicz et al. 2019). Ideally, plants prefer ammonium over nitrate, because ammonium does not need to be reduced before incorporation into plant compounds. In soils with good aeration, oxidation of ammonium is rapid and, as a consequence, nitrate is generally present in higher concentrations in soil than is ammonium. Furthermore, the relative ease of movement of nitrate through the soil accelerates its absorption by plants. Consequently, most plants have developed to grow better with nitrate and, a number of studies have shown that plant growth may be enhanced with a mixed supply of ammonium and nitrate. Although there are some exceptions, for example about rice, nitrate is not stable in submerged soils since nitrate is lost during the denitrification process and therefore ammonium must be supplied instead of nitrate (Hofman and Van Cleemput, 2004).

N content in dry plant tissue is generally between 1 to 4 percent, and this content is higher in leguminous plants. Among the different parts of plant structure which have nitrogen, protein N is the largest N fraction. Nitrogen requirements vary from less than 100 kg N/ha.year to more than 400 kg N/ha.year, depending on the nitrogen content and production of different parts of the plant (Hofman and Van Cleemput, 2004).

Attention to environmental and economic issues has increased the need to better understand the role and fate of N in crop production systems. Nitrogen deficiency as a nutrient is seen in crop production all over the world and its use can bring significant economic returns for farmers. The stress caused by the lack of nutrients in plants can significantly reduce agricultural performance. Nitrogen, an essential nutrient, is a critical growth-limiting factor and is a major component of proteins, DNA/RNA, and chlorophyll. Plant traits such as plant height, number of leaves, surface and color, etc. are under the effects of N deficiency. The first symptoms of N deficiency are usually slow growth and uniform yellowing of older leaves. The whole plant looks pale to yellowish green and early senescence of older leaves can be seen. Nitrogen-deficient plants produce smaller fruits, leaves, and shoots than normal, and they can develop more slowly than normal. Broadleaf foliage in fall may be more reddish than normal and drop prematurely. Moreover, increased root growth and stunted shoot growth results in a low shoot/root ratio (Marschner et al., 2023).

N deficiency decreases crop growth and production. Visual detections and diagnostics are valuable tools for assessing the nutritional status of crops. However, if visual symptoms are observed, it means the plants are suffering from and affected by that stress, which can lead to reduced yields. Visual diagnosis of deficiencies needs experience and can only successfully be accomplished by experts (Hofman and Van Cleemput, 2004).

A shortage of available N is characterized by stunted plants, less than optimum growth rate, and the older leaves senesce prematurely. N deficiency results in leaf chlorosis, sometimes with distinctive patterns. N deficiencies first appear on older leaves since, during a deficiency, nitrogen from the older leaves will be metabolized and transported to the new ones. Nitrogen deficiency reduces the synthesis of amino acids and, consequently, of proteins, resulting in reduced growth and accumulation of non-nitrogen metabolites, and makes photosimilates more accessible for use in the synthesis of secondary metabolism compounds, ascorbic acid, among other organic acids (Marschner, 2023; Hofman and Van Cleemput, 2004). Nevertheless, it should be kept in mind that visual detection of deficiency symptoms is only to guide the occurrence of possible nutrition-related problems, since several nutrients are responsible for the color formation in the leaves of plants. Iron (Fe) and magnesium (Mg) are also responsible for the synthesis of chlorophyll in leaves. Hence, the successful use of leaf color as an index is dependent on eliminating, minimizing, or detection of all factors other than N that can also affect leaf color (Vásquez-Jiménez and Bartholomew, 2018).

Biological Nitrogen Fixation (BNF)

Biological nitrogen fixation denotes to a microbial process based on the enzyme "nitrogenase" converting atmospheric nitrogen into root-absorbable ammonium. Nitrogen-fixing microorganisms, collectively referred to as "diazotrophs," are able to biologically fix atmospheric nitrogen in association with plant roots. The BNF process can be broadly divided into two groups: symbiotic and non-symbiotic or free-living. The symbiotic group is those that fix nitrogen in symbiosis with the host plant, and the free-living group is independent of the host plant, but the support of the host will add to their efficiency. For instance, symbiotic associations of rhizobia and legume are known to be the most important BNF biosystem, contributing with an average of 200-300 kg N/ha.yr (Herridge et al., 2008). While, non-symbiotic BNF estimates for non-legume (e.g., maize, rice and wheat) production systems based on a 50-year assessment study reported an average contribution of 20–30 kg N/ha.yr. Estimates of BNF rates by the non-symbiotic method vary due to several factors including environmental variation, management and cultivation practices, genotypic differences, and technical features related to the methods used to predict BNF (Peoples and Herridge, 2000). Nitrogenase is an enzyme encoded by the *nif* gene. The presence of this enzyme in *Azotobacter* makes the most characteristic of *Azotobacter* to be formed and these soil bacteria are qualified as biofertilizers. This enzyme is responsible for converting N_2 to NH_3. Nitrogenase is a complex enzyme and contains two enzymes, the dinitrogenase (an iron-molybdenum protein which is also known as component I) and the dinitrogenase reductase (an iron protein which is also known as component II) (Pankievicz et al., 2019). This enzyme for its activity, in addition to atmospheric nitrogen, needs a strong reductant, ATP, and an anaerobic environment to maintain catalytic activity.

The electron transfer from dinitrogenase reductase to dinitrogenase needs the intermediation of Mg-ATP. The reaction of nitrogen fixation by nitrogenase is shown below. According to this reaction, under ideal conditions, 8 electrons and at least 16 MgATP are required to convert 1 mole of N_2 to 2 moles of NH_3. This equation can be considered with a little difference for HUP- and HUP+, but this reaction applies to the HUP- group (Paul, 2007):

$$N_2 + 8e + 16MgATP + 8H \text{-----------} 2NH_3 + H_2 + 16MgADP + 16Pi$$

Most diazotrophs require anaerobic conditions for nitrogen fixation. Not only the nitrogenase enzyme is deactivated by oxygen in diazotrophs, but the genes involved in nitrogen fixation are also suppressed by oxygen. This is while *Azotobacter* fixes nitrogen under aerobic conditions. This genus has one of the highest respiratory quotients among all biological systems examined and its "respiratory protection" has been accepted as a theory to explain oxygen tolerance of nitrogen fixation in Azotobacters (Das, 2019). Therefore due to high content of respiration rate in Azotobacteria, oxygen present inside the cells would be consumed at a very rapid rate resulting in very low intracellular oxygen concentration.

So far, numerous bacterial species were identified to have nitrogen fixing properties in non-symbiotic manner, including *Azotobacter* sp., *Azospirillum* sp., *Beijerinckia* sp., *Pseudomonas* sp., *Herbaspirillum*, *Gluconacetobacter*, *Burkholderia*, *Clostridium*, *Methanosarcina*, *Bacillus*, *Paenibacillus*, etc. (Kizilkaya, 2008; Deaker et al., 2011).

A Sustainable Approach to Increase the Bioavailability of Iron in Plants   167

Free-Living Nitrogen Fixing Bacteria

One of the most important properties of beneficial bacteria related to the improvement of plant growth and/or health is the feature of nitrogen fixation ability. N_2-fixers are classified into two categories. The first category includes root/legume-associated symbiotic bacteria which possess specificity and infect the roots to produce nodule e.g., strains of *Rhizobium* and generally known as *Rhizobia*. Another group of bacteria is the so-called free-living N_2-fixing bacteria, which have no plant specificity (Oberson et al., 2013). Although, in some literature they are classified into four groups: symbiotic nodule-formers, rhizospheric associative, soil free-living and endophytic (Paul, 2007). *Azospirillum*, *Azotobacter*, *Burkholderia*, *Herbaspirillum*, *Bacillus*, and *Paenibacillus* are examples of such free-living nitrogen fixers (Goswami et al., 2016).

Azotobacteria are a group of free-living nitrogen fixing bacteria, heterotrophic, non-spore forming and Gram negative, and aerobic bacteria inhabiting the soil, some of them are motile by means of peritrichous flagella while others are immotile. These bacteria are variable in terms of morphology, cells range from straight rods with rounded ends to more ellipsoidal or coccoid, depending on the age and culture medium, moreover under unfavorable environmental conditions form thick-walled cysts (dormant cells resistant to deleterious conditions). Polymorphic has been reported in this genus (for example some species such as *A. paspali* can be filamentous) and it ranges in size from 2 to 10 mm in length and 1 to 2 mm in width (Garrity, 2007).

Almost seven species in the genus *Azotobacter* have been reported (Martyniuk and Martyniuk, 2003). This bacterium was recognized as the first aerobic free nitrogen fixer in 1901 by the Dutch microbiologist, botanist and founder of environmental microbiology-Beijerinck and his colleagues. *Azotobacter* spp. use N_2 for the synthesis of their cellular protein, which is mineralized in the soil and gives plants a significant part of the nitrogen available from the soil source. These bacteria are sensitive to acidic pH, high salt concentration and temperature (Aquilanti et al., 2004). The cyst formation is a process in which a single vegetative cell vegetative to become spherical shape which is known as cyst (Ebrahimi et al., 2020), encystment occurs during late stationary phase and is induced naturally in face of unfavorable and extreme conditions such as high or low temperatures, freezing, salinity, and drought, also in response to changes in nutrients concentrations in the medium. Cyst formation has been observed in some species (Garrity, 2007).

Azotobacteria are well known because of some traits such as the highest respiratory rates among living organisms, the production of capsules and slime in the presence of carbohydrates in solid media, the production of pigment and cyst formation (Ebrahimi et al., 2017). Pigment production is a common feature in this genus and some species produce yellow-green, red-violet, or brownish-black pigments. Among *Azotobacter* species, *A. chroococcum* is the dominant species and it usually appeared in brown-black non diffusible pigment in aging colonies. Production of insoluble black-brown pigment in water is the characteristic of this species which is dominant in agricultural soils in temperate zones (Aquilanti et al., 2004; Garrity, 2007).

Diversity of Azotobacteria

Azotobacteria are common in soils sampled from across the world. This genus along with *Pseudomonas* family belonges to the Gamma-proteobacter and comprises seven species containing *Azotobacter chroococcum*, *A. vinelandii*, *A. beijerinckii*, *A. salinestri*, *A. paspali*, *A. nigricans* and *A. armeniacus* (Garrity, 2007). In other resource such as German collection of microorganisms and cell cultures (DSMZ) 12 species of Azotobacters have been reported (https://lpsn.dsmz.de/genus/azotobacter).

The properties and diversity of Azotobacteria may be influenced by factors such as the age of the plants and the type of plant grown, soil type and elevation, agricultural practices applied, composition the microbial communities present, root exudates and chemical compounds applied to the soil (Ebrahimi et al., 2017). There are reports on the genetic diversity of indigenous diazotrophic bacteria, and the ability of free-living NFB to increase crop productivity under different environmental conditions (soil, water, and air) has already been revealed (Tejera et al., 2005).

Frequency and distribution of Azotobacteria are influenced by chemical (soil organic matter, soil acidity and salinity), physical (soil temperature, soil depth, soil moisture) and biological properties of soil (microbial interactions) and land usage. Since they need P for BNF, they are more commonly found in fertile soils than in non-fertile soils such as sandy soils (Garrity, 2007). Furthermore, land usage and land management affect the distribution of microbial population (Ebrahimi et al., 2017). Azotobacteria are normally found in slightly acidic to alkaline soils. Generally, lower pH (<6.0) decreases *Azotobacter* population and in some cases, completely stops their growth. An optimum pH of 7–7.5 is satisfactory for the physiological functions of

A Sustainable Approach to Increase the Bioavailability of Iron in Plants   169

Azotobacter. For example, *A. beijerinckii* species were often reported in acidic soils while *A. chroococcum* and *A. vinelandii* species are more abundant in tropical soils. Meanwhile, the species *A. paspali* is only specifically associated with the plant roots of *Paspalum notatum* cv *Batatais* (Kennedy et al., 2015). Optimum temperature for growth and physiological properties of *Azotobacter* is 25–30ºC, because this genus is a mesophilic organism. Azotobacters cannot tolerate high temperatures, although they can survive at 45-48°C by forming cysts that germinate under favorable conditions.

The number of *Azotobacter* genus in soils is generally reported low ($<10^4$ CFU/g soil). However, they are found throughout the world typically in 30 to 80% of sampled soils (Kennedy et al., 2015). Due to its importance in agricultural and non-agricultural soil a research was performed by Ebrahimi et al. (2017) to study the occurrence and population of *Azotobacter* in some soil samples with different features and applications. The ability of neural network model as an alternative for estimating population of *Azotobacter* from physicochemical and biological properties of soil data was focus of this research. They reported that using 5 variables of soil as inputs of the model, including (soil texture; e.g., sand and silt), soil salinity (EC), substrate induced respiration (SIR), and content of carbonate calcium equivalent (CCE) in soil, this parameter can be estimated. According to the findings of this study factors influencing on soil *Azotobacter* number were EC, pH, OC, soil texture and carbonate calcium.

To obtain an overview of *Azotobacter* diversity, we used nucleotide sequence data filed on the NCBI website up to May 19, 2023. Using keywords such as "Azotobacter" and "Azotobacter 16S rRNA," 7413 and 4086 reports were found, respectively. Checking these data showed some misdata which after controlling them the final data belonged to gamma-proteobacteria reduced to 6853 and 3990, respectively. In our investigation, we believe that "Azotobacter 16S rRNA" keyword results are reliable to consider the diversity of this genus. Therefore, to demonstrate its diversity, we only used this keyword. These data are certainly increasing as research in this area is ongoing around the world. It should be noted that few of these final reports (98 out of 3990) belonged to other genera e.g., (*Azomonas, Pseudomonas, Klebsiella,* etc.) and were ignored in this study. Our data analysis showed that almost 42.16% of identified Azotobacters belonged to species *A. chroococcum*, while *A. beijerinckii* (32.58%) and *A. vinelandii* (19.14%) were in the next order. As shown in Figure 2, *A. salinestris* (1.64%), *A. tropicalis* (1.46%), had lower than 2% contribution in this diversity while for species including *A. armeniacus* (0.33%) and *A. nigricans* (0.33%) this contribution was less than

0.5%. Some of this diversity belonged to the uncultured bacterium (0.77%) and Azotobacters without any identified species which known as *Azotobacter* sp. (1.56%).

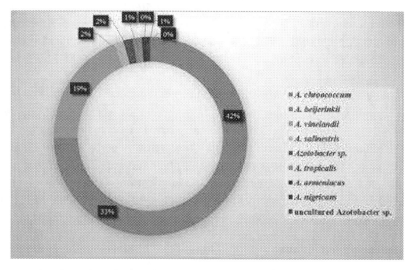

Figure 2. Diversity of *Azotobacter* genus. These data extracted from NCBI website to show the relative distribution of different species of *Azotobacter* genus.

Other Mechanisms of Action of *Azotobacter*

N fixation is considered as an interesting microbial activity on the earth's surface as it provides a way of recycling the nitrogen and plays an important role in nitrogen homeostasis in the biosphere (Paul, 2007). In addition, as an important biological process, BNF also contributes to maintaining soil fertility and increasing crop productivity (Goswami et al., 2016). *Azotobacter* is able to convert atmospheric nitrogen to ammonia, which is then taken up and used by the plants (Das, 2019). There are reports indicate that *Azotobacter* has the potential of fixing N about 20 kg/ha.yr and therefore it can be successfully used in crop production as a substitute for at least part of nitrogen mineral fertilizers (Kizilkaya, 2008). There are reports that plants inoculated with *Azotobacter* required less nitrogen fertilizer. For example, it was reported that the application of a consortia of *Azotobacter* spp. could reduce the need of N-fertilizers up to fifty percent (Romero-Perdomo et al., 2017).

Plant growth-promoting bacteria (PGPB) are a group of beneficial bacteria which improve directly or indirectly plant growth and/or health.

A Sustainable Approach to Increase the Bioavailability of Iron in Plants   171

These bacteria are present in the rhizosphere, on the rhizoplane and associated with roots. Several mechanisms such as phytohormone production, nitrogen fixation, synthesis of siderophore, solubilization of minerals (P, K, Fe, Zn, etc) and control of disease are identified and reported for these bacteria in various studies (Goswami et al., 2016; Sarikhani et al., 2018; 2019; 2020). So far, different mechanism of action has been reported for improved growth and health of inoculated plants with *Azotobacter* sp. However, the exact mode of action behind the growth promoting activity of *Azotobacter* is not fully explored although a very rich literature regarding the use of *Azotobacter* in plant growth promotion is available yet. The key characteristics of *Azotobacter* sp. action can be nitrogen fixation, production of phytohormone like Indole-3-Acetic Acid (IAA), solubilization of P or production of siderophore. There are some reports that shows these diazotrophs are capable to solubilize insoluble P and K forms in the soil (Ebrahimi et al., 2018; 2020; 2023). Hormones such as (auxins, cytokinins, and GA-like substances) not only enhance plant growth and nutrient uptake, but can also indirectly protect host plants from phytopathogens and stimulate other beneficial rhizosphere microorganisms (Goswami et al., 2016).

Isolation and Screening of Azotobacteria

Until now, just a few number of the identified diazotrophic species have found their way into agriculture and used as a commercial biofertilizer. PGPB especially Azotobacteria are considered as important biofertilizers, since they pave the way to reduce application of chemical fertilizers in modern agriculture and represent an alternative to manage organic and low-input agriculture. Because of the adaptation of indigenous bacteria to ecological conditions, they have advantages for such applications. Therefore, it seems that isolation of native Azotobacters from dissimilar environments and characterization of their plant growth-promoting properties is a promising method in the development of commercial biofertilizers (Ebrahimi et al., 2023). According to above mentioned parts, isolation and screening process of these beneficial bacteria is important and is considered in this section.

Different isolation strategies have been used in diverse studies such as using nitrogen free (NF) media (e.g., Ashby, Winogradsky, Beijerinckia, LG, Brown, Burk and Jensen) with or without enrichment process. In order to isolate Azotobacteria, in addition to use the soil paste, usually NF mediums are being used. In this method, in addition to performing serial dilutions of soil samples, the final dilutions are plated on NF medium and soft, flat, clear, or milky slimy colonies are identified after incubation for at least 3 days, and are counted as *Azotobacter* (Figure 3) (Ebrahimi et al., 2018; 2022; 2023).

The pre-enrichment process is usually carried out in the NF solution for 7-14 days, followed by streaking onto the solid NF medium. Alongside these two methods, using soil paste is another approach to isolate Azotobacters. Approximately 50 g of soil sample should be accurately mixed with 0.5% (w/w) sodium pyruvate while adding sterile water to the mixture. The soil paste, prepared in a bigger glass petridish, should be transferred and pressed inside a petri dish with a sterile spatula to obtain a smooth and levelled surface. After proper time of incubation (3–7 days) at 27–30 °C, the soil paste–plates presenting growth of *Azotobacter* were revealed by the appearance of slimy, glistening colonies, turning brown with aging if produced by the species *A. chroococcum* (Figure 4).

Because it is time-consuming and expensive to test large numbers of bacteria in greenhouses or field studies, the generally accepted strategy is to reduce the number of bacteria by testing promising bacteria in in-vitro studies. Therefore, first their potential from this point of views (especially nitrogen fixing ability, P solubility, K releasing ability, synthesize of auxin, production of siderophore) should be investigated and the robust ones selected for pot culture and field trials (Ebrahimi et al., 2023).

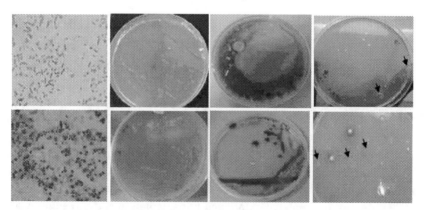

Figure 3. Process to obtain pure colonies of nitrogen fixing bacteria using either LG (upper line) or Winogradsky medium (lower line) for screening. a.) and e.) diluted soil extract, b.) and f.) streak culture during the purification process on NF media, c.) and g.) purified bacterial isolate (streak culture) on nutrient agar, d.) vegetative form of a pure bacterial isolate on NA after Gram staining, h.) cyst form of a pure bacterial isolate after Gram staining (Ebrahimi et al., 2018).

Our experience in the isolation and purification of *Azotobacter* has shown that these bacteria are usually not purified so easily, and usually during the purification process in the subcultures, other related bacteria grow below and near the *Azotobacter* colony. This problem was reported by Farajzadeh et al. (2012). Such contaminations has also been observed around the cysts in the old cultures. They reported and showed this problem by Gram staining of the unique colonies of *Azotobacter*. They stated that the majority of the isolates were accompanied with one or more other bacterial species and their molecular identification revealed that this non-target bacteria or contamination belong to *Achromobacter* genus. Another main problem of this genus is their storing and re-culturing. Usually, after storing these bacteria in glycerol, making a new culture from this bacterium fails.

Figure 4. Isolation and purification process of *Azotobacter* from soil samples, growth and pigment formation of *Azotobacter* on soil paste and LG media and gram staining and observing of cyst (from right to left) (Ebrahimi et al., 2018).

Role of Azotobacteria in Sustainable Agriculture

For more than a century, Azotobacters have been used as biofertilizers. In addition to fixing nitrogen, these bacteria are able to produce plant hormones and while dissolving various elements (such as phosphorus, etc.), they also suppress plant pathogens or reduce their harmful effects. Application of native and wild type Azotobacters has been reported to result in better yield of cereals (e.g., corn, wheat, oat, barley, rice, and sorghum), of oil seeds (e.g., mustard and sunflower), of vegetable crops (e.g., tomato, eggplant, carrot, chillies, onion, potato, beans and sugar beet), of fruits (e.g., mango and sugar cane) and of tree like oak (Sumbul et al., 2020; Aasfar et al., 2021).

The first commercial biofertilizer that was produced and launched in the world was "Nitrogin," a *Rhizobium*-based bioinoculant. Further study of soil N_2-fixing microorganisms (including non-symbiotic bacteria such as *Azotobacter*) led to the development of another biofertilizer in Russia and

Eastern European countries. This *Azotobacter*-based biofertilizer was known as "Azotobacterin." Various formulations of Azotobacters alone and of combinations with other microbes are available commercially for use as biofertilizers. Nowadays, the use of this bacterium as a biofertilizer has received more attention all over the world, and for example, "Azotobarvar" (containing *A. vinelandii*) is a brand of this biofertilizer that has recently been produced and used in Iran (Sarikhani and Amini, 2020). Using biofertilizer alone or in combination with some fertilizers or pesticides gives good results in some research papers. Since *Azotobacter* is a non-symbiotic microbe, its maximum potential for increasing plant productivity can be achieved by inoculating it simultaneously with some other biofertilizers compared to its single application.

In a study, Moradi and Sarikhani (not published data) isolated *Azotobacter* strians from the soil samples of Botanical Garden, University of Tabriz, Iran Their potential were investigated in in-vitro conditions (such as the ability of the growth in NF media, P solubility and K releasing activity, auxin production and so on) and they were evaluated in pot culture in comparison with negative (without bacterial inoculation and fertilizer application) and positive control (chemical fertilizer) treatments. The results of this study showed a positive and significant effect of inoculation of *Azotobacter* strain (3MDP-4) on both maize shoots and roots (Figure 5).

Figure 5. Inoculation of corn with *Azotobacter* sp. (3MDP-4) strains in a pot culture. (From left to right, negative control, positive control and 3MDP-4).

Conclusion

To discover the potential of *Azotobacter* in soil fertility, more research is necessary in the future. Azotobacters are being used as biofertilizer for more than one century, since 1902. These bacteria are regarded as PGPB which besides biological nitrogen fixation (an average of 20 kg N/ha.year), can exude phytohormones (IAA, gibberellic acid, and cytokinins), can solubilize phosphates, release potassium and can counteract plant pathogens. The beneficial effect of inoculation of Azotobacteria on various plants such as cereals, oilseeds, fruits, vegetables and trees has been reported. This bacteria helps to increase the availability of nutrients and restore soil fertility for better crop response. Due to its important role in soil sustainability, it is an important component of the integrated nutrient management system. However, this type of biofertilizer cannot fully match chemical fertilizers in terms of increasing yield and competition. Therefore, some chemical fertilizer should be added along with biofertilizer to achieve optimal yield. The problem is that nitrogen fixation by nitrogen fixing bacteria especially Azotobacteria is completely suppressed by N chemical fertilizers. All this type of nitrogenous fertilizers produce ammonia and this product inhibits *nif* gene expression.

Azotobacteria are one of the beneficial microbes that can be used as biofertilizers to produce environmentally friendly and sustainable products. However, there is an urgent need for further studies related to improving screening techniques, isolation and identification of plant growth promoting compounds and antimicrobial compounds from bacterial isolates and clarifying the molecular basis of the mechanisms involved. Advances in culture-independent methods, mainly amplicon sequencing and shotgun, can significantly improve our understanding of the diversity and performance of diazotrophs, including *Azotobacter* species, in soil and plant compartments. A challenge for the research community is to find compatible partners, i.e., a particular strain of Azotobacteria that associates well with a particular plant genotype, to ensure maximum benefit from biofertilizers. The formulation of microbial inoculants still requires fundamental and applied studies, paying attention to the new formulation of liquid or carrier based inoculants or the production of encapsulated form of this bacterium, and its use as a consortia with other beneficial bacteria can be new era of research on Azotobacters.

References

Aasfar A, Bargaz A, Yaakoubi K, Hilali A, Bennis I, Zeroual Y and Meftah Kadmiri I. Nitrogen Fixing Azotobacter Species as Potential Soil Biological Enhancers for Crop Nutrition and Yield Stability. *Front. Microbiol.* (2021) 12:628379.

Anas, Muhammad, Fen Liao, Krishan K. Verma, Muhammad Aqeel Sarwar, Aamir Mahmood, Zhong-Liang Chen, Qiang Li, Xu-Peng Zeng, Yang Liu, and Yang-Rui Li. "Fate of nitrogen in agriculture and environment: agronomic, eco-physiological and molecular approaches to improve nitrogen use efficiency." *Biological Research* 53, no. 1 (2020): 1-20.

Aquilanti, L., F. Favilli, and F. Clementi. "Comparison of different strategies for isolation and preliminary identification of Azotobacter from soil samples." *Soil Biology and biochemistry* 36, no. 9 (2004): 1475-1483.

Benghzial, Kaoutar, Hind Raki, Sami Bamansour, Mouad Elhamdi, Yahya Aalaila, and Diego H. Peluffo-Ordóñez. "GHG global emission prediction of synthetic N fertilizers using expectile regression techniques." *Atmosphere* 14, no. 2 (2023): 283.

Bindraban, Prem S., Christian Dimkpa, Latha Nagarajan, Amit Roy, and Rudy Rabbinge. "Revisiting fertilisers and fertilisation strategies for improved nutrient uptake by plants." *Biology and Fertility of Soils* 51, no. 8 (2015): 897-911.

Brahmaprakash, G. P., and Pramod Kumar Sahu. "Biofertilizers for sustainability." *Journal of the Indian Institute of Science* 92, no. 1 (2012): 37-62.

Das, Hirendra Kumar. "Azotobacters as biofertilizer." *Advances in applied microbiology* 108 (2019): 1-43.

Deaker, Rosalind, M. László Kecskés, M. Timothy Rose, K. Amprayn, Krishnen Ganisan, Thi Kim Cue Tran, Thuy Nga Vu, Thi Cong Phan, Thanh Hien Nguyen, and I. Robert Kennedy. *Practical methods for the quality control of inoculant biofertilisers.* Australian Centre for International Agricultural Research (ACIAR), 2011.

Ebrahimi, Mitra, Ali Akbar Safari Sinegani, Mohammad Reza Sarikhani, Seyed Abolghasem Mohammadi, Nasser Aliasgharzad, and Ralf Greiner. "Plant Growth-Promoting Traits and Genetic Diversity of Free-Living Nitrogen-Fixing Bacteria Isolated from Soils in North of Iran." *Iranian Journal of Science and Technology, Transactions A: Science* 46, no. 3 (2022): 761-769.

Ebrahimi, Mitra, Ali Akbar Safari Sinegani, Mohammad Reza Sarikhani, and Seyed Abolghasem Mohammadi. "Comparison of artificial neural network and multivariate regression models for prediction of Azotobacteria population in soil under different land uses." *Computers and Electronics in Agriculture* 140 (2017): 409-421.

Ebrahimi, Mitra, Ali Akbar Safari Sinegani, Mohammad Reza Sarikhani, and Seyed Abolghasem Mohammadi, and Nasser Aliasgharzad. Isolation, identification, and determination of plant growth promoting properties of Azotobacteria isolated from soil samples north-west of Iran under different land use. *Applied Soil Research* (2018): 27-42.

Ebrahimi, Mitra, Ali Akbar Safari Sinegani, Mohammad Reza Sarikhani, and Nasser Aliasgharzad. "Inoculation effects of isolated plant growth promoting bacteria on wheat yield and grain N content." *Journal of Plant Nutrition* 46, no. 7 (2023): 1407-1420.

A Sustainable Approach to Increase the Bioavailability of Iron in Plants 177

Farajzadeh, Davoud, Bagher Yakhchali, Naser Aliasgharzad, Nemat Sokhandan-Bashir, and Malak Farajzadeh. "Plant growth promoting characterization of indigenous Azotobacteria isolated from soils in Iran." *Current microbiology* 64 (2012): 397-403.

Garrity, George. *Bergey's Manual® of Systematic Bacteriology: Volume 2: The Proteobacteria, Part B: The Gammaproteobacteria.* Vol. 2. Springer Science & Business Media, 2007.

Goswami, Dweipayan, Janki N. Thakker, and Pinakin C. Dhandhukia. "Portraying mechanics of plant growth promoting rhizobacteria (PGPR): a review." *Cogent Food & Agriculture* 2, no. 1 (2016): 1127500.

Herridge, David F., Mark B. Peoples, and Robert M. Boddey. "Global inputs of biological nitrogen fixation in agricultural systems." *Plant and soil* 311 (2008): 1-18.

Hofman, Georges, and Oswald Van Cleemput. "Soil and plant nitrogen." (2004).

Kennedy, C., Rudnick, P., MacDonald, M. L., Melton, T. (2015). *"Azotobacter," in Bergey's Manual of Systematics of Archaea and Bacteria,* eds. M. E. Trujillo, S. Dedysh, P. DeVos, B. Hedlund, P. Kämpfer, F. A. Rainey and W. B. Whitman (Atlanta, GA: American Cancer Society), 1–33. doi: 10.1002/9781118 960608.gbm01207.

Kızılkaya, Rıdvan. "Yield response and nitrogen concentrations of spring wheat (Triticum aestivum) inoculated with Azotobacter chroococcum strains." *Ecological Engineering* 33, no. 2 (2008): 150-156.

Kumar, Satish, Satyavir S. Sindhu, and Rakesh Kumar. "Biofertilizers: An ecofriendly technology for nutrient recycling and environmental sustainability." *Current Research in Microbial Sciences* 3 (2022): 100094.

Marschner, Petra, and Zed Rengel. "Nutrient availability in soils." In *Marschner's Mineral Nutrition of Plants,* pp. 499-522. Academic press, 2023.

Martyniuk, S., and M. Martyniuk. "Occurrence of Azotobacter spp. in some Polish soils." *Polish Journal of Environmental Studies* 12, no. 3 (2003): 371-374.

Oberson, Astrid, Emmanuel Frossard, C. Bühlmann, J. Mayer, P. Mäder, and A. Lüscher. "Nitrogen fixation and transfer in grass-clover leys under organic and conventional cropping systems." *Plant and Soil* 371 (2013): 237-255.

Pankievicz, Vânia, Thomas B. Irving, Lucas GS Maia, and Jean-Michel Ané. "Are we there yet? The long walk towards the development of efficient symbiotic associations between nitrogen-fixing bacteria and non-leguminous crops." *BMC biology* 17, no. 1 (2019): 1-17.

Paul, E. A. "Soil microbiology, ecology, and biochemistry in perspective." In *Soil microbiology, ecology and biochemistry,* pp. 3-24. Academic Press, 2007.

Peoples, Mark B., and David F. Herridge. "Quantification of biological nitrogen fixation in agricultural systems." *Nitrogen fixation: from molecules to crop productivity* (2000): 519-524.

Romero-Perdomo, Felipe, Jorge Abril, Mauricio Camelo, Andrés Moreno-Galván, Iván Pastrana, Daniel Rojas-Tapias, and Ruth Bonilla. "Azotobacter chroococcum as a potentially useful bacterial biofertilizer for cotton (Gossypium hirsutum): Effect in reducing N fertilization." *Revista Argentina de microbiologia* 49, no. 4 (2017): 377-383.

Sarikhani, M. R., Sh Oustan, M. Ebrahimi, and N. Aliasgharzad. "Isolation and identification of potassium-releasing bacteria in soil and assessment of their ability to release potassium for plants." *European journal of soil science* 69, no. 6 (2018): 1078-1086.

Sarikhani, Mohammad Reza, and Rohollah Amini. "Biofertilizer in sustainable agriculture: Review on the researches of biofertilizers in Iran." *Journal of Agricultural Science and Sustainable Production* (2020).

Sarikhani, Mohammad Reza, Nasser Aliasgharzad, and Bahman Khoshru. "P solubilizing potential of some plant growth promoting bacteria used as ingredient in phosphatic biofertilizers with emphasis on growth promotion of Zea mays L." *Geomicrobiology Journal* 37, no. 4 (2020): 327-335.

Sarikhani, Mohammad Reza, Bahman Khoshru, and Ralf Greiner. "Isolation and identification of temperature tolerant phosphate solubilizing bacteria as a potential microbial fertilizer." *World Journal of Microbiology and Biotechnology* 35 (2019): 1-10.

Sumbul A, Ansari RA, Rizvi R, Mahmood I. Azotobacter: A potential bio-fertilizer for soil and plant health management. Saudi Journal of Biological Sciences 27 (2020) 3634–3640.

Tejera, N. L. C. M. M. G. J., C. Lluch, M. V. Martinez-Toledo, and J. Gonzalez-Lopez. "Isolation and characterization of Azotobacter and Azospirillum strains from the sugarcane rhizosphere." *Plant and soil* 270 (2005): 223-232.

Vásquez-Jiménez, Jhonny, and Duane P. Bartholomew. "Plant nutrition." In *The pineapple: botany, production and uses*, pp. 175-202. Wallingford UK: CAB Interna.

Chapter 8

The Role of Biofertilizers in Sustainable Agriculture

Bhupinder Dhir

School of Sciences, Indira Gandhi National Open University, New Delhi, India

Abstract

Biofertilizers are comprised of microbial formulations that promote growth of plants by facilitating uptake/assimilation of nutrients. Biofertilizers increase growth of plants via processes such as nitrogen fixation, solubilization of phosphorus, uptake of water, and synthesis of growth-promoting substances. Increase in the growth of the crop plants leads to increase in agricultural output. Indirectly biofertilizers support plant growth by protecting them from environmental stresses (such as salinity, drought stress, high temperature) and biotic agents (such as pathogens). Bioengineering techniques can be used to enhance the efficiency of microbial communities that form an important component of biofertilizers. Low nutritional value, availability of a suitable carrier and use in in high amounts restrict the use of biofertilizers at a large scale. Research studies related to the long-term effects of biofertilizers on health of living organisms and various components of environment need to be assessed. Biofertilizers, thus can prove to be a sustainable, environment-friendly alternate to the harmful chemical fertilizers.

Keywords: environment, biofertilizer, agriculture, soil fertility, sustainable

In: Biofertilizers
Editor: Philip L. Bevis
ISBN: 979-8-89113-082-1
© 2023 Nova Science Publishers, Inc.

Introduction

Chemical fertilizers and pesticides have been used since years to increase crop productivity. Extensive use of agrochemicals results in environmental deterioration and produces various kinds of health hazards in living organisms (Sharma and Singhvi, 2017). Input of excess of the chemical fertilizers deteriorate soil quality. This happens because of change in physical and chemical properties. Many alternate materials have been explored in the last few years with an aim of improving soil quality and restoration of its fertility. Biofertilizers have shown their potential as an eco-friendly and sustainable alternate to the harmful chemical fertilizers (Mitter et al., 2021).

The microorganisms that form an important component of biofertilizerscolonize rhizosphere or the interior of the plants (Raja, 2013; Bhattacharjee and Dey, 2014; Patel et al., 2014; Kumar and Singh, 2015). The microbial communities help in improving the quality of the soil and restoration of its fertility (Ritika and Uptal, 2014). Biofertilizers facilitate uptake of mineral nutrients by plants and also increase the accessibility of nutrients to plants (Singh et al., 2015a; Simarmata et al., 2016). They assist in mobilization of nutrients such as nitrogen, phosphorus, sulphur via nitrogen fixation, phosphorus solubilization, oxidation of sulphur or decomposition of organic matter. Since biofertilizers assist in providing continuousl supply of nutrients nin the soil, they have become an important component of Integrated Nutrient Management (INM) system (Adesemoye and Kloepper, 2009).

Mode of Application of Biofertilizers

Biofertilizers formulation mainly includes inoculum, glycerol (a cell protectant), carrier material and a packaging material. Lignite is a carrier that is commonly used in the formulation. It is rich in organic matter and is capable of holding more water which promotes the growth of microorganisms.

Biofertilizers can be applied to soil in solid form or as a liquid suspension. The biofertilizers are mainly applied in the form of powder formulations, dry mixture and adhesive slurry. All these forms can be applied to seeds. Fertilizers coated with material like lime are also applied to soil (Bashan, 1998). Besides the solid form of biofertilizer, water soluble form of these fertilizers is also used. Liquid biofertilizers consist of suspension of microorganisms and cell protectants or chemicals that enhance formation of

The Role of Biofertilizers in Sustainable Agriculture 181

latent spores or cysts. Liquid biofertilizers remain stable for a longer duration (upto 2 years) and show tolerance to unfavourable environment such as high temperature (Hegde, 2008; Ansari et al., 2015). These forms are able to maintain high cell count for longer time duration (Verma et al., 2011). The advantages such as easy handling, easy application and long storage have made biofertilizers popular among farmers (Herrmann and Lesueur, 2013; Allouzi et al., 2022).

The dry form of biofertilizer is applied directly to soil. . Granules (size 1–2 mm) made from tank bed clay (TBC) are soaked overnight in a suspension of bacteria and air-dried at room temperature followed by heating at 200°C in a muffle furnace. This process increases porosity of the granules and helps in their sterilization. These granules can be applied to fields along with seeds.

Types of Bio-Fertilizers

Bio-fertilizers are classified on the basis of the group of microorganisms they contain or the type of element they supplement in the soil (Table 1). Bio-fertilizers are mainly classified as follows:

Table 1. Different types of biofertilizers (Barman et al., 2017)

Type of Biofertilizer	Examples
Nitrogen fixing bacteria	*Azotobacter, Clostridium, Anabaena, Nostoc, Rhizobium, Azospirillum*
Phosphorus solubilizing bacteria	*Bacillus subtilis, Pseudomonas striata*
Potassium solubilizing bacteria	*Frateuria aurantia*
Micronutrient solubilizing Bacteria (Silicate and Zinc solubilizers)	*Bacillus* sp.

Nitrogen Biofertilizers

Many microbial species present in the soil fix atmospheric nitrogen and convert them to organic forms. These microbes are generally found associated with the nodules present in root of plants, particularly legumes. These microbes convert atmospheric nitrogen into organic form of nitrogen which can be easily used by the plants. Hence, they improve the availability of nitrogen to the plants. *Rhizobium* sp.*, Azospirillum* sp,*. Azotobacter,*

Bejerinkia, Clostridium, Klebsiella, Anabaena, Nostoc and blue-green algae. are some of the important nitrogen fixing species.

Rhizobium forms symbiotic association with roots of legume plants (Jehangir et al., 2017). Legumes show nitrogen fixation capacity of about 50-100 kg N/ha after symbiotic association. Rhizobium association has been found in legume crops like chickpea, red-gram, pea, lentil, black gram, soybean and groundnut (Table 2). The *nod, nif* genes present in the bacterium are responsible for regulating nodulation and nitrogen fixation.

Azospirillum shows nitrogen fixing capacity of 20-40 kg/ha. It is used in non-leguminous crop species like sugarcane, sorghum, maize, pearl millet. *Azospirillum* species mainly *A. amazonense, A. halopraeferens* and *A. brasilense* help in nitrogen fixation. Biologically active substances such as vitamins, nicotinic acid, indole acetic acids and gibberellins produced by *Azospirillium* help in promoting plant growth.

Azotobacter is a free living non-symbiotic nitrogen fixing bacteria. It is found associated within the rhizosphere of crop plants such as rice, maize, sugarcane, bajra, vegetables and many plantation crops (Wani et al., 2013). It show nitrogen fixation capacity of 15-20 kg/ha N per year. It releases substances such as vitamins, thiamine and riboflavin, nicotinic acid, pantothenic acid, biotin, plant-growth hormones which promote the growth of plants. Anti-fungal compounds produced by *Azotobacter* inhibit growth of several pathogenic fungi. *Azotobacter chroococcum* is the most common species while other species include *A. vinelandii, A. beijerinckii, A. insignis* and *A. macrocytogenes* that assist in nitrogen fixation.

Acetobacter is another nitrogen-fixing bacteria and is generally found associated with sugar-producing crops. The growth hormones produced by it support growth of roots and shoot. Improvement in root growth and density increases mineral and water uptake from the soil, thus supporting plant growth.

Cyanobacteria are free-living single celled or multi-celled symbiotic blue green algae. They fix nitrogen, mobilize insoluble phosphate and improve physical and chemical properties (such as aeration and water holding capacity) of soil. They produce growth-promoting substances such as auxin.

Azolla helps in nitrogen fixation and is mainly used as a green manure. Increase in yield of plants after use of *Azolla* has been reported.

Other free-living nitrogen fixing bacteria include *Clostridium pasteurinnum,* photosynthetic bacteria such as *Rhodobacter, Azotobacter* and *Methanogens.*

The Role of Biofertilizers in Sustainable Agriculture   183

Table 2. *Rhizobium* species and their host plants
(Ponmurugan and Gopi, 2006)

Rhizobium species	Host Legume
R. leguminosarum	Pea, sweet pea
R. meliloti	Sweet clover
R. trifoli	Clover / berseem
R. phaseoli	All beans
Bradyrhizobium japonicum	Lupins

Phosphate Biofertilizers

Bacteria present in the soil and rhizospheric zone convert insoluble phosphorus into soluble phosphorus making them available to plants. The mineralization of organic phosphorus compounds is facilitated by enzymes phosphatase and phytase. *Rhizobium* increase the availability of phosphorus via solubilization. Organic acids produced by bacteria solubilize phosphorus by lowering pH of the soil and helping in dissolution of bound forms of phosphate (Chang and Yang, 2009). Lowering of pH helps in exchange of phosphate ions by acid ions. *Flavobacterium, Pseudomonas, Mycobacterium, Achromobacter, Aereobacter, Rhizobium, Burkholderia, Microccocus, Erwinia* are some of the microbial species that act as phosphate solubilizers. Bacterial species solubilize insoluble inorganic phosphate present in compounds such as tricalcium phosphate, dicalcium phosphate, hydroxyapatite and rock phosphate.

Arbuscular mycorrhizae fungal species such as *Glomus, Gigaspora, Acaulospora Scutellospora, Sclerocystis* and some ectomycorrhizal species such as *Laccaria, Pisolithus, Boletus and Amanita* increase mobility of phosphorus in soil (Ghorbanian et al., 2012; Ewa et al., 2013).

Potassium Biofertilizers

Microbes including bacteria, fungi and actinomycetes solubilize potassium present in the soil. Inorganic and organic acids produced by these microbes assist in the solubilization of potassium. Availability of potassium in soil increases due to processes such as acidolysis, chelation, and exchange reaction (Archana et al., 2013; Meena et al., 2015). Bacterial species including *Bacillus licheniformis, B. mucilaginosus, B. edaphicus, B. circulans, Pseudomonas azotoformans, Burkholderia, Acidithiobacillus ferrooxidans*, *Enterobacter hormoechei* show shown high potassium solubilizing ability (Prajapati et al., 2013; Saha et al., 2016). Some phosphate solubilizing bacteria such as *Bacillus* sp. also show capacity to mobilize potassium.

Zinc Biofertilizers

Microorganisms also help in solubilization of zinc (Mahdi et al., 2010). Bacterial species that assist in zinc solubilization include *Bacillus subtilis, Thiobacillus thioxidans* and *Saccharomyces* sp. The organic acids produced by bacteria solubilize silicates and helps in the release of metal ions thereby making them available to plants.

Sulphur Oxidizing Biofertilizer

Bacteria such as *Thiobacillus* show capacity to oxidize sulphur to form sulfates which can easily be used by plants.

Plant Growth Promoting Rhizobacteria (PGPR)

The group of bacteria that colonize roots or rhizosphere soil and promote the growth of plants are referred to as plant growth promoting rhizobacteria (PGPR) (Vessey, 2003; Radzki and Vessey, 2003; Verma et al., 2010; Gupta et al., 2015; Rochlani et al., 2022). Major PGPR species include *Bacillus spp.* and *Pseudomonas fluorescence.*

PGPR alter physical and chemical properties of the soil (Glick, 2012). Most of the PGPR species stimulate proton-pump ATPase which enhance mineral absorption by plants. Ion flux at the surface of roots gets increased and this plays a role in promoting uptake of nutrient by plants. Organic acids produced by these bacteria decrease pH of the soil. The decrease in pH help in chelation of divalent cations such as the Ca, Zn, Mg and phosphates.

Exopolysaccharide (EPS) produced by PGPR adhere to surface of soil and help in maintaining high moisture content in the dry soil. EPS form a protective sheath or biofilm around the roots of the plants which protects them from desiccation. Hence, EPS assist in survival of plants under conditions of

The Role of Biofertilizers in Sustainable Agriculture

abiotic stress. Exopolysaccharide (EPS) producing PGPR species include *Rhizobium sp., Enterobacter cloacae, Bacillus pretences, Azotobacter vinelandi. Pseudomonas putida* in particular produce siderophores that increase the availability of iron.

PGPR species synthesize hormones which enhance plant growth. Growth regulators such as indole-acetic acid, cytokinins, gibberellins produced by them promote the growth of roots by improving their absorptive capacity for water and nutrients. Studies indicate that *Azospirillum* secrete gibberellins, ethylene and auxins, while *Rhizobium* and *Bacillus* synthesize IAA in large amounts. In contrast, PGPR also enhance the growth of plant by suppressing the synthesis of ethylene.

PGPR promote growth of plants by acting as bio-protectants. PGPR produce lytic enzymes and metabolites that regulate various pathogenic agents (Khosro and Yousef, 2012; Bhattacharyya and Jha, 2012). PGPR species such as *Pseudomonas* show defense against various plant diseases. Chitinase, SOD (superoxide dismutase), and ACC (1-aminocyclopropane-1-carboxylic acid) deaminase are some of the enzymes produced by PGPR that kill pathogens via damage to cell wall. Enzymes such as beta 1-3 glucanase and chitinase rupture cell wall and restrict fungal development. Studies have shown that *Pseudomonas fluorescence* LPK2 produce enzymes such as chitinase and beta-glucanases that control wilt producing fungus namely *Fusarium oxysporum* and *Fusarium udum*. PGPR induce resistance in plants via salicylic acid-dependent systematic acquired resistance (SAR) pathway. Antibiotics produced by *Pseudomonas* and *Bacillus* produce antagonistic effects and trigger induced systematic resistance (ISR).

PGPR provide resistance against abiotic stress by creating ROS scavenging system and improving leaf water status under abiotic stress conditions.

Mycorrhiza

Vesicular Arbuscular Mycorrhiza (VAM) act as broad-spectrum biofertilizers. External hyphae of mycorrhiza help in holding the soil particles together and lead to formation of micro-aggregates which later get converted into macro-aggregate. They improve accumulation of plant nutrients (Khosro and Yousef, 2012; Wani et al., 2013). They stimulate physiological activities in plants and/or reduce severity of diseases caused by soil pathogens.

Mechanism of Action of the Biofertilizers

Biofertilizers promote growth and development of plants. They enhance plant growth by producing tolerance against biotic and abiotic stresses and providing nutrition by fixing atmospheric nitrogen and solubilizing soil nutrients. Increase in plant growth occurs because of direct and indirect mechanisms (Figure 1).

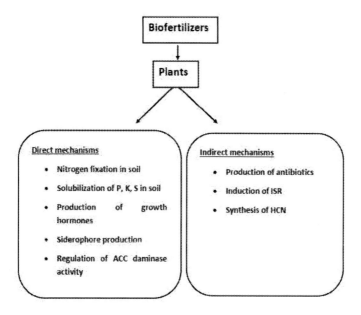

Figure 1. Role of biofertilizers in plant growth via direct and indirect mechanisms.

The direct mechanisms by which biofertilizers improve plant growth include

a) nitrogen fixation
b) solubilization of phosphate via production of organic acids
c) production of phytohormones
d) procurement of iron via siderophores
e) production of exopolysaccharides

Indirectly biofertilizers promote growth by reducing the impact of pathogenic microorganisms. They suppress pathogens and other harmful microorganisms through parasitism, producing antagonistic substances such

The Role of Biofertilizers in Sustainable Agriculture   187

as antibiotics, hydrogen cyanide, siderophores and enzymes. They induce systemic resistance against a wide range of pathogenic microbes. Siderophores, EPS and cyanides produced by biofertilizers act as plant pathogen antagonists and protect plants from pathogen attack. Lytic enzymes such as glucanases, proteases, chitinases and antibiotics play a role in enhancing resistance capacity of the host. Hydrogen cyanide (HCN) produced by *Bacillus* and *Pseudomonas* acts as an effective biocontrol agent and shows the capacity to kill cells by interfering with processes electron transport chain (ETC) and energy supply. PGPR produce enzyme 1-aminocyclopropane-1-carboxylate (ACC) deaminase that regulate the level of ethylene in plants (Singh et al., 2015b).

Functions of Biofertilizers

Improvement in Quality of Soil

Biofertilizers improve quality of the soil. They increase the nutrient availability in the soil by facilitating nutrient absorption by the roots and improvement in nutrient exchange capacity of soil. Biofertilizers increase the water holding capacity of the soil and improve moisture content of soils. Biofertilizers improve soil structure and physical conditions of the soil by bringing improvement in structure and aggregation of soil particles, reducing compaction, increasing pore capacity and water infiltration. They increase soil aeration and water percolation. They stabilize the C: N ratio of the soil.

Improvement in Plant Growth

Biofertilizers promote plant growth and increase yield by increasing the availability of nutrients via mechanisms such as nitrogen fixation, solubilization and mobilization of nutrients. They promote development of roots. The hormones and vitamins produced by them help in plant growth. IAA produced by PGPR induce production of nitric oxide (NO), which acts as a second messenger and triggers complex signaling network leading to improvement in growth of roots and developmental processes. PGPR strains enhance plant growth by suppressing expression of ethylene. The synthesis of ethylene from 1-aminocyclopropane-1-carboxylate (ACC), an immediate precursor of ethylene is inhibited due to hydrolysis by enzyme bacterial ACC-deaminase (Glick, 2014). Improvement in the hydraulic conductivity of roots helps in uptake of water by plants.

Improvement in Plant Metabolic Activities

Certain strains of *Rhizobia* increase surface areas of plant leaves which leads to increase in net photosynthetic rate, stomatal conductance and water use efficiency. Studies have indicated increase in photosynthetic capacity of the rice after inoculation with *Rhizobium* (Mia and Shamsuddin 2010). Use of three bacterial biofertilizers namely *Pseudomonades, Bacillus lentus* and *A. brasilense* in combination increased chlorophyll content in leaves of plants exposed to stress (Heidari and Golpayegani 2012). Biofertilizer, thus show the capacity to promote photosynthetic activity of the plant and helps in plant growth in stressed conditions.

Protection against Abiotic Stress

Abiotic and biotic factors affect the productivity of crops. Application of biofertilizers produced a positive effect and showed stimulation in the plant growth under stress conditions. Mycorrhizal and bacterial association improve the productivity of crops exposed to drought and salinity stress (Singh, 2020).

Studies have shown that PGPR exposure improve leaf water status in plants exposed to abiotic stress. Use of PGPR reduces the level of stress-inducing hormone ethylene in plants via synthesis of enzyme 1-aminocyclopropane-1- carboxylate (ACC) deaminase (Singh et al., 2015). Plants inoculated with arbuscular mycorrhiza fungi showed improvement in growth under salt stress (Ansari et al., 2019). Inoculation of PGPR alone or along with AM such as *Glomus intraradices, G. mosseae* showed high nutrient uptake and improvement in physiological processes in plants exposed to stress conditions. Inoculation of *Rhizobium trifoli* along with *Trifolium alexandrium* increased biomass production and nodulation in plants exposed to salinity stress (Hussain et al., 2022).

Inoculation of *Pseudomonas putida* RS-198 increased rate of seed germination and seedling growth in cotton exposed to alkaline and high salt conditions. The growth parameters including plant height, fresh weight and dry weight showed increase in these plants. Uptake of elements like potassium, magnesium, calcium increased while absorption of sodium decreased in such plants. Osmolytes and proteins produced by *P. fluorescens* MSP-393 helped in overcoming negative effects of salt stress. Plant treated with *P. mendocina* showed increase in shoot biomass under salt stress conditions. Hence *Pseudomonas* species specifically proved beneficial in reducing the effects of salt stress in plants.

Inoculation of arbuscular mycorrhiza along with nitrogen fixing bacteria help in overcoming the effects of drought stress in legumes. Application of *Pseudomonas* sp. improved anti-oxidant defense machinery and enhanced production of photosynthetic pigment in plants exposed to water stress. Rice plant inoculated with arbuscular mycorrhiza showed an increase in photosynthetic efficiency and anti-oxidative response under drought conditions (Ruiz-Sanchez et al., 2011). Inoculation with *Pseudomonas* sp. exerted positive effect on the seed germination, seedling growth, improve synthesis of antioxidant and photosynthetic pigments in water stressed plants (Heidari and Golpayegani, 2012).

Some biofertilizers provide tolerance to plant against more than one abiotic stress. Inoculation of *Achromobacter piechaudii* enhanced growth and increased biomass production in tomato and pepper plants exposed to salt and water stress. Calcisol produced by PGPRs such as *Pseudomonas alcaligenes*, *Bacillus polymyxa* and *Mycobacterium phlei* helped in overcoming stress caused due to high temperature and salinity. Treatment of *Pseudomonas putida* or *Bacillus megaterium* along with AM fungi proved effective in alleviating drought stress. Application of *Pseudomonas, Bacillus lentus and A. brasilense* increased chlorophyll content and synthesis of antioxidant enzymes in leaves of plants exposed to stress (Heidari and Golpayegani, 2012).

Protection against Biotic Stresses

PGPRs provide resistance against pathogen attacks. Bacteria, mycorrhizae confer resistance against various pathogens. They have been found effective against pathogens causing diseases such as spotted wilt viruses in tomato, cucumber mosaic virus of tomato and pepper and banana bunchy top virus in banana. Several *Bacillus* species have been found to provide resistance against various pathogens. Application of *Bacillus amyloliquefaciens* 937b and *B.*

pumilus SE-34 have shown capacity to provide immunity against tomato mottle virus. *B. megaterium* IISRBP 17 prove effective against *Phytophthor* and *Fusarium* infestation in banana. *Bacillus subtilis* GBO3 showed potential to provide resistance against *Rhizoctonia solani* in cotton plants. *Paenibacillus polymyxa* proved effective against *Fusarium* infection in watermelon. The metabolites produced by biofertilizers were found responsible for providing resistance against pathogens (Backman and Sikora, 2008; Saraf et al., 2014).

Arbuscular mycorrhizal fungi (AMF) reduce damage caused by fungi, nematodes and bacteria. Mycorrhizal colonization reduced disease outbreak in plants. Many root pathogens such as *Rhizoctonia solani, Pythium spp., Fusarium oxysporum* showed inhibition after AM inoculation. These plants showed improvement in plant nutrient profile and productivity. *Glomus mosseae* provided resistance against *Fusarium oxysporum* in basil plants. Treatment of arbuscular mycorrhizal fungi and *Pseudomonas fluorescens* reduced the development of root-rot disease but at the same time reduced yield of *Phaseolus vulgaris.*

Biofertilizers reduced the efficiency to reduce parasitic nematodes (Youssef and Eissa, 2014). Application of *Azospirillum* and *Azotobacter* increased growth and yield in chilli infected with nematode (Khan et al., 2012). *Paenibacillus polymyxa, Bacillus megaterium* and *Bacillus circulans* showed high nematicidal activity against the root-knot nematode in *Meloidogyne incognita* (El-Haddad et al., 2011).

Remediation of Pesticides

Azospirillum, Azotobacter, Actinomycetes, Bacillus, Enterobacter, Gordonia, Klebsiella, Paenibacillus, Pseudomonas, Serratia are some of the microbes that show capacity to degrade and transform pesticides thereby reducing their toxicity in plants (Shaheen and Sundari, 2013). The breakdown of the compounds occurs via enzymatic degradation. The degradation is catalyzed by enzymes such as hydrolases, esterases, mixed function oxidases (MFO) and glutathione S-transferases (GST) (Ortiz-Hernández et al., 2013). The breakdown of the compounds occurs via processes such as hydrolysis, oxidation, dehalogenation, addition, reduction, replacementand cleavage (Ramakrishnan et al., 2011).

Providing Resistance against Metal Stress

Bacterial species viz. Achromobacter xylosoxidans, A. chroococcum, Bacillus subtilis, B. megaterium, Bradyrhizobium, Pseudomonas sp., Brevibacillus sp., Kluyvera ascorbata, Mesorhizobium, Pseudomonas putida, Pseudomonas aeruginosa, Ralstonia metallidurans, Rhizobium, Sinorhizobium sp., Variovox paradoxus, Ochrobactrum sp., Psycrobacter sp., and Xanthomonas sp. show potential to reduce stress caused by heavy metals (Shinwari et al., 2015). PGPR decrease metal toxicity via production of microbial siderophores (Radzki et al., 2013). Siderophores form stable complexes with toxic metals such as Cd, Cu, Cr, Pb, Zn thereby reducing stress imposed by metals (Dimkpa et al., 2009; Saha et al., 2016). The metal toxicity can also be reduced by biosorption via microbial cells also reduce metal toxicity viamicrobial cells. Pseudomonas putida, and P. fluorescens have shown protection from inhibitory effects of cadmium via IAA, siderophore and 1-aminocyclopropane-1-carboxylate deaminase (ACCD) synthesis in canola and barley plants.

Reclamation of Degraded Land

Biofertilizers also help in restoration of land degraded by acidic soil due to mining activities. Biofertilizers help in restoring the fertility and nutrient level of damaged soil. Mycorrhizal associations have shown potential to restore degraded soils of mining sites (Mukhopadhyay and Meiti, 2009; Quoreshi et al., 2019). Studies have shown that *Azotobacter, Rhizobium,* phosphate solubilizing microbes, blue green algae and VAM combination helped in restoring silicate mining area.

Effect of Biofertilizers on Agricultural Productivity

Biofertilizers application improve yield of crop plants (Aseri et al., 2008; Gharib et al., 2008; Isfahani and Besharati, 2012; Bhardwaj et al., 2014; Ranjan et al., 2020). Studies have shown that application of biofertilizers increase yield of crop plants by 20–30%. Improvement in soil fertility, increase in nutrient status support plant growth and this leads to increase in yield of plants (Table 3). Application of *Rhizobium* biofertilizers has shown enhancement in agronomic yield of pulse crops. Improvement in leaf area index, harvest index and yield attributes were noted in many crop plants after *Azospirillum* application (Singh et al., 2016). The grain yield increased by 10%-40% in plants treated with biofertilizers (Kumar et al., 2018). Inoculation of *Azolla* increasedavailability of nitrogen in rice plants (Yadav et al., 2019).

Table 3. Bio-fertilizer recommended for various crops

Biofertilizer	Crops
Rhizobium	Pulses, Oilseeds, Fodders
Azospirillum	Rice, Wheat, millets, maize, sorghum, sugarcane
Azotobacter	Rice, Wheat, millets, other cereals, cotton, vegetables, sunflower, mustard, flowers
Azolla	Rice

Table 4. Biofertilizers and their effects on crop plants

Plant	Type of biofertilizer	Effect of biofertilizer	Reference
Cymbopogon martinii	*Glomus aggregatum, Bacillus polymyxa* and *Azospirillum brasilense*	Increase in biomass and phosphorus content	Ratti et al., 2001
Melons	Arbuscular mycorrhizae (AM)	Total yield	Padilla et al., 2006
Amaranthus	*Azospirillum* and *Pseudomonas*	Increase in biomass production and growth	Thenmozhi et al., 2010
Zea mays	*Azospirillum* spp. and *Pseudomonas fluorescens*, alone or in combination	high N and P content in tissue	Sangoquiza et al., 2019
Zea mays	Cyanobacteria consortium	Increase in plant height, number of leaves	Hernández-Reyes et al., 2019
chickpea	combined application of phosphate solubilizing bacteria and *Trichoderma harzianum*	Increase in leaf and grain P content	Mohammadi et al., 201
Zea mays	*Herbaspirillum seropedicae*	Increase in weight and length of ear, number of grains, cob weight and grain yield	Avila et al., 2020
Potato	*Enterobacter cloacae*	Enhanced growth and yield, significant increase in root and shoot length and biomass	Verma and Agarwal., 2018

The Role of Biofertilizers in Sustainable Agriculture 193

Treatment of arbuscular mycorrhizal (AM) fungi also increased growth in crop plants. The yield of crop plants increased by about 10% after treatment with AM. Combination of AM along with nitrogen-fixing bacteria or compost extracts also brought increase in yield of crop plants including rice.

Significant increase in yield of pulse crops was noted after application of *Rhizobium* biofertilizers. Application of *Azospirillum* improved growth of agricultural crops and this was shown by increase in parameters such as leaf area index, harvest index, and yield attributes (Singh et al., 2016).

Application of commercial biofertilizer with the name of Vital N® that contained *Azospirillium* induced growth in roots of crop plants like corn, rice, banana, garlic, orchids and onion. *Azospirillium* produces plant-growth-stimulating substance indole-3-acetic acid (IAA) that promote growth and increase yield.

Advantages of Biofertilizers

Use of biofertilizers has got wider acceptance because of the positive effects exerted on soil and growth of plants. Some of the advantages has been listed below:

- They act as good alternate of harmful chemical fertilizers. They can replace chemical fertilizers by 25%-30%.
- They are cost effective in comparison to commercial chemical fertilizers.
- They are eco-friendly and does not pollute the environment (Kumar et al., 2022).
- Their handling and application is easy.
- They form an important component of the integrated nutrient supply system.
- They help in biocontrol of pathogens. Siderophores and antibiotics produced by fungal biofertilizers are antagonistic to foliar or rhizosphere pathogenic bacteria, fungi and insects.
- They exert positive effect on soil microbiology via control of phytopathogenic organisms.
- They act as bio-ameliorators and promote plant growth under stressed conditions.

Reasons That Restrict the Use of Biofertilizers

Biofertilizers, though offer a wide range of advantages to soil and plants but possess certain features or properties that restrict their application. Some of the adverse effects of use of biofertilizers include

- Short shelf life
- Requirement of a suitable carrier for maintaining shelf life. Broth cultures do not have carrier protection and lose their viability quickly (Herrmann and Lesueur, 2013).
- They show susceptibility to high temperature
- Storage of biofertilizers- These fertilizers need to be stored at room temperature, and not below 0°C nor above 35 °C (Chen, 2006).
- Lack of awareness among farmers- Knowledge and skills related to the correct application of biofertilizers is also lacking among farmers.
- Low level of acceptance by farmers.
- Variable efficiency of microbes- Efficiency of microbes according to change in environmental conditions. The changes in pH, moisture, temperature and other environmental variables affect the growth and viability of microbes.
- The microbial strains should be able to survive in both broth and inoculant carriers.
- Antagonistic microorganisms present in the soil and non-specific host–inoculant relationships can prove threat to the survival of microbial communities. . Microflora and fauna compete with the inoculated microorganisms for nutrients and other vital factors in the micro-ecological niches.
- The lack of adequate formulations restrict the use of biofertilizers.
- Bio-fertilizers need to be used carefully. They need to be protected from sunlight and should not be mixed with nitrogen fertilizers or applied along with fungicides.
- Variable amount of nutrients may be present or macronutrients may not be present in sufficient quantities in biofertilizers, hence large volumes are required.
- Over use of fertilizers can cause risks to soil and living organisms.
- Environmental issues–

The Role of Biofertilizers in Sustainable Agriculture 195

Extensive and long-term application can lead to change in soil properties (physicochemical and biological). Accumulation of salts, nutrients (such as nitrogen, ammonia) and heavy metals in soil affect soil and/or water quality, hence cause significant reduction of plant growth (Buragohain et al., 2017; Yadav et al., 2018).

- Marketing strategies for increasing sale of biofertilizers are missing. Support from government authorities for the promotion of application of biofertilizers for sustainable agriculture are required.

Conclusion and Recommendations

Biofertilizers are cost-effective alternate to harmful chemical fertilizers. They play an important role in improving growth and yield of plants. They promote growth of plants via interactions in the rhizospheric zone and stimulating processes such as nitrogen fixation, phosphorus solubilization which help in release of nutrients in the soil. They synthesize growth-promoting substances which also promote growth in plants. Thus, use of biofertilizers can be a a good approach for sustainable agriculture. Research studies are required to increase the efficiency of these fertilizers for various soil types. Studies related to assessment of environmental changes after the long-term and overuse of biofertilizers are required. Awareness programmes related to production of inoculants and their application are required. Awareness program and training sessions focusing on use of biofertilizers need to be conducted for their large-scale implementation.

References

Adesemoye AO, Torbert HA, Kloepper JW. Plant Growth-Promoting Rhizobacteria Allow Reduced Application Rates of Chemical Fertilizers. *Microbial Ecology* (2009) 58: 921-929.
Allouzi MMA, Allouzi SMA, Keng ZX, Supramaniam CV, Singh A, Chong S. Liquid biofertilizers as a sustainable solution for agriculture. *Heliyon* (2022) 8(12):e12609.
Ansari M, Shekari F, Mohammadi MH, Juhos K, Végvári G, Biró B. Salt-tolerant plant growth-promoting bacteria enhanced salinity tolerance of salt-tolerant alfalfa (*Medicago sativa* L.) cultivars at high salinity. *Acta Physiologiae Plantarum* (2019) 41: Article number: 195.

Ansari MF, Tipre DR, Dave SR. Efficiency evaluation of commercial liquid biofertilizers for growth of *Cicer aeritinum* (chickpea) in pot and field study. *Biocatalysis and Agricultural Biotechnology* (2015) 4: 17-24.

Archana DS, Nandish MS., VP. Savalagi AA. Characterization of potassium solubilizing bacteria (KSB) from rhizosphere soil. *BIOINFOLET* (2013) 10: 248- 257.

Aseri GK, Jain N, Panwar J, Rao AV, Meghwal PR. Bio-fertilizer s improve plant growth, fruit yield, nutrition, metabolism and rhizosphere enzyme activities of Pomegranate (*Punica granatum* L.) in Indian Thar Desert. *Scientia Horticulturae* (2008) 117(2): 130-135.

Ávila JS, Ferreira JS, Santos JS, Rocha PAD, Baldani VL. Green manure, seed inoculation with *Herbaspirillum seropedicae* and nitrogen fertilization on maize yield. *Rev Bras Eng Agrícola Ambient (*2020) 24:590–595.

Backman PA, Sikora RA. Endophytes: an emerging tool for biological control. *Biol Control* (2008) 46: 1-3.

Barman M, Paul S, Choudhury AG, Roy P, Sen J. Biofertilizer as Prospective Input for Sustainable Agriculture in India. *Int J Curr Microbiol App Sci* (2017) 6(11): 1177-1186.

Bashan Y. Inoculants of plant growth-promoting bacteria for use in agriculture. *Biotechnol Adv* (1998) 16(4): 729–770.

Bhardwaj D, Ansari MW, Sahoo RK, Tuteja N. Biofertilizers function as key player in sustainable agriculture by improving soil fertility, plant tolerance and crop productivity. *Microbial Cell Factories* (2014) 13:66.

Bhattacharjee R, Dey U. Biofertilizer, a way toward organic agriculture: a review. African *Journal of Microbiol Research* (2014) 8(24):2332–2342.

Bhattacharyya PN, Jha DK. Plant growth-promoting rhizobacteria (PGPR): emergence in agriculture. *World Journal of Microbiotechnology* (2012) 28: 1327 – 1350.

Buragohain S, Sarma B, Nath DJ, Gogoi N, Meena RS, Lal R. Effect of 10 years of biofertiliser use on soil quality and rice yield on an Inceptisol in Assam. *India Soil Research* (2017) 56(1): 49-58.

Chang CH, Yang SS. Thermo-tolerant phosphate solubilizing microbes for multi-functional biofertilizer preparation. *Bioresour Technol* (2009) 100:1648-1658.

Chen J. The Combined Use of Chemical and Organic Fertilizers and/or Biofertilizer for Crop Growth and Soil Fertility. International Workshop on Sustained Management of the Soil-Rhizosphere System for Efficient Crop Production and Fertilizer Use, Bangkok, (2006) 1-11.

Dimkpa CO, Merten D, Svatoš A, Büchel G, Kothe E. Metal induced oxidative stress impacting plant growth in contaminated soil is alleviated by microbial siderophores. *Soil Biology and Biochemistry* (2009) 41(1):154–162.

El-Haddad ME, Mustafa MI, Selim SM, El Tayeb TS, Mahgoob AE, Aziz NH. The nematicidal effect of some bacterial biofertilizers on Meloidogyne incognita in sandy soil. *Braz J Microbiol* (2011) 42(1):105–113.

Ewak, Ewa O, Piotr S, Anna S, Jolanta JS. Effect of Pseudomonas luteola on mobilization of phosphorus and growth of young apple trees (Ligol)-Pot experiment. *Scientia Horticulturae* (2013) 164:270-276.

Gamalero E, Glick BR. Bacterial ACC deaminase and IAA: interactions and consequences for plant growth in polluted environments. Handbook of Phytoremediation (2010) 763–774.

Gharib FA, Moussa LA, Massoud ON. Effect of compost and bio-fertilizer s on growth, yield and essential oil of sweet marjoram (*Majorana hortensis*) plant. *International Journal of Agriculture and Biology* (2008) 10(4): 381-387.

Ghorbanian D, Harutyunyan S, Mazaheri D, Rasoli V, MohebI A. Influence of Arbuscular mycorrhizal fungi and different levels of phosphorus on the growth of corn in water stress conditions. *African J Agric Res* (2012) 7(16):2575-2580.

Glick BR. Plant growth promoting bacteria: mechanisms and applications. *Scientifica* (2012) 2012:15.

Glick BR. Bacteria with ACC deaminase can promote plant growth and help to feed the world. *Microbiol Research* (2014) 169(1):30–39.

Gupta G, Parihar SS, Ahirwar NK, Snehi SK, Singh V Plant growth promoting rhizobacteria (PGPR): current and future prospects for development of sustainable agriculture. *Journal of Microbial and Biochemical Technology* (2015) 7:2.

Hegde SV Liquid bio-fertilizers in Indian agriculture. *Bio-fertilizer newsletter* (2008) p. 17-22.

Heidari M, Golpayegani A. Effects of water stress and inoculation with plant growth promoting rhizobacteria (PGPR) on antioxidant status and photosynthetic pigments in basil (*Ocimum basilicum* L.). *J Saudi Soc Agric Sci* (2012) 11(1):57–61.

Hernández-Melchor DJ, Ferrera-Cerrato R, Alarcón A. Trichoderma: Importancia agrícola, biotecnológica, y sistemas de fermentación para producir biomasa y enzimas de interés industrial. *Chilean Journal of Agricultural & Animal Sciences* (2019) 35:98-112.

Herrmann L, Lesueur D. Challenges of formulation and quality of biofertilizers for successful inoculation. *Appl Microbiol Biotechnol* (2013) 97(20): 8859–8873.

Hussain MB, Shah S, Matloob A, Shareef MN, Mubaraka R. Rice Interactions with Plant Growth Promoting Rhizobacteria In book: *Modern Techniques of Rice Crop Production* (2022), pp. 231-255.

Isfahani FM, Besharati H. Effect of biofertilizers on yield and yield components of cucumber. *J Biol Earth Sci* (2012) 2:83-92.

Jehangir IA, Mir MA, Bhat MA, Ahangar MA. Biofertilizers an approach to sustainability in agriculture: A review. *Int J Pure App. Biosc.* (2017) 5(5): 327-334.

Khan Z, Tiyagi SA, Mahmood I, Rizvi R. Effects of N fertilization, organic matter, and biofertilizers on the growth and yield of chilli in relation to management of plant-parasitic nematodes. *Turk J Bot* (2012) 36(1): 73–78.

Khosro M, Yousef S (2012). Bacterial bio-fertilizers for sustainable crop production: A review. *APRN Journal of Agricultural and Biological Science* 7 (5): 237 – 308.

Kumar MS, Reddy GC, Phogat M, Korav S. Role of bio-fertilizers towards sustainable agricultural development: A review. *Journal of Pharmacognosy and Phytochemistry* (2018) 7(6): 1915-1921.

Kumar S, Satyavir D, Sindhu S., Kumar R. Biofertilizers: An ecofriendly technology for nutrient recycling and environmental sustainability. *Current Research in Microbial Sciences* (2022) 3: 100094.

Kumar S, Singh A. Biopesticides: present status and the future prospects. *Journal of Biofertilizer Biopesticides* (2015) 6: e129.

Mahdi SS, Hassan GI, Samoon SA, Rather HA, Dar SA, Zehra B. Bio-fertilizers in organic agriculture. *Journal of Phytology* (2010) 2:42-54.

Meena RK, Singh R K, Singh NP, Meena SK, Meena VS. Isolation of low temperature surviving plant growth – promoting rhizobacteria (PGPR) from pea (*Pisum sativum* L.) and documentation of their plant growth promoting traits. *Biocatalysis and Agricultural Biotechnology* (2015) 4: 806-811.

Mia MB, Shamsuddin ZH. Nitrogen fixation and transportation by rhizobacteria: a scenario of rice and banana. *International Journal of Botany* (2010) 6:235–242.

Mitter EK, Tosi M, Obregón D, Dunfield KE, Germida JJ. Rethinking Crop Nutrition in Times of Modern Microbiology: Innovative Biofertilizer Technologies. *Front Sustain Food Syst* (2021) 5: article 606815.

Mohammadi K, Ghalavand A, Aghaalikhani M, Heidari GR, Sohrabi Y. Introducing the sustainable soil fertility system for chickpea (*Cicer arietinum* L.). *African Journal of Biotechnology* (2011) 10(32): 6011-6020.

Mukhopadhyay S, Meiti SK. Biofertiliser: VAM fungi- A future prospect for biological reclamation of mine degraded lands. *Indian Journal of Environmental Protection* (2009) 29: 801-808.

Ortiz-Hernández ML, Sánchez-Salinas E, Dantán-González E, CastrejónGodínez ML. Pesticide biodegradation: mechanisms, genetics and strategies to enhance the process. Biodegradation-life science. Intech-publishing, Rijeka, (2013) pp. 251–287.

Padilla E, Esqueda M, Sánchez A, Troncoso-Rojas R, Sánchez A. Effect of biofertilizers on cantaloupe crop with plastic mulching. *Rev Fitotec Mex* (2006) 29 (4): 321 – 329.

Patel N, Patel Y, Mankad A. Bio fertilizer: a promising tool for sustainable farming. *Int J Innov Res Sci Eng Technol* (2014) 3:15838-15842.

Ponmurugan P, Gopi C. *In vitro* production of growth regulators and phosphatase activity by phosphate solubilizing bacteria. *African Journal of Virology Research* (2006) 12 (1): 001-003.

Prajapati K, Sharma MC, Modi HA. Growth promoting effect of potassium solubilizing microorganisms on okra (*Abelmoschus esculentus*). *International Journal of Agricultural Science and Research* (2013) 3 (1): 181-188.

Quoreshi AM, Suleiman MK, Manuvel AJ, Sivadasan MT, Jacob S, Thomas R. Biofertilizers for Agriculture and Reclamation of Disturbed Lands: An Eco-friendly Resource for Plant Nutrition. International Conference on Applied Science, Technology and Engineering *Journal of Mechanical Contamination & Mathematical Science* (2019) Science Special Issue (4): 231-243.

Rachael AC, Gospel AC, Nelson, Kalu CT, Gabriel C, Nwosu, Ulunma O, Egbufor, Christian U, Godswill AC. A Review on the Role of Biofertilizers in Reducing Soil Pollution and Increasing Soil Nutrients. *Himalayan Journal of Agriculture* (2020) 1: 34-38.

Radzki W, Manero FG, Algar E, García JL, García-Villaraco A, Solano BR. Bacterial siderophores efficiently provide iron to iron starved tomato plants in hydroponics culture. *Antonie Van Leeuwenhoek* (2013) 104(3):321–330.

Radzki W, Vessey JK (2003) Plant growth promoting rhizobacteria as biofertilizers. *Plant Soil* (2020) 255(2): 571–586.

The Role of Biofertilizers in Sustainable Agriculture 199

Raja N. Biopesticides and biofertilizers: Eco-friendly sources for sustainable agriculture. *Journal of Biofertilizer Biopesticide* (2013) (3): 112-115.

Ramakrishnan B, Megharaj M, Venkateswarlu K, Sethunathan N, Naidu R. Mixtures of environmental pollutants: effects on microorganisms and their activities in soils. In Reviews of Environmental Contamination and Toxicology, Springer New York (2011) 211:63–120.

Ranjan S, Sow S, Choudhury SR, Kumar S, Ghosh M. Biofertilizer as a Novel Tool for Enhancing Soil Fertility and Crop Productivity: A Review. *Int J Curr Microbiol App Sci* Special Issue (202) 211: 653-665.

Ratti, N, Kumar, A Vermin, HN Gantains SP. Improvement in bioavailability of tricalcium phosphate by *Rhizobacteria* and *Azospirillum* inoculation. *Microbiology Research* (2001) 156: 145-149.

Ritika B, Uptal D. Bio-fertilizer a way towards organic agriculture: A Review. *Academic Journals* (2014) 8(24):2332-42.

Rochlani A, Dalwani A, Shaikh NB, Shaikh N, Sharma S, Saraf SM. Plant Growth Promoting Rhizobacteria as Biofertilizers: Application in agricultural sustainability. *Acta Scientific Microbiology* (2022) 5(4): 12-21.

Ruíz-Sánchez M, Armada E, Munoz Y, de Salamonec IEG, Aroca R, Ruíz-Lozano JM, Azcónb R. *Azospirillum* and arbuscular mycorrhizal colonization enhance rice growth and physiological traits under well-watered and drought conditions. *Journal of Plant Physiology* (2011) 168: 1031–1037.

Saha M, Maurya BR, Meena VS, Bahadur I, Kumar. Identification and characterization of potassium solubilizing bacteria (KSB) from Indo-Gangetic Plains of India. *Biocatalysis and Agricultural Biotechnology* (2016) 7: 202-209.

Sangoquiza C, Yanez C, Borges M. *Respuesta de la absorción de nitrógeno y fósforo de una variedad de maíz al inocular Azospirillum sp. y Pseudomonas fluorescens*. ACI Avances En Ciencias e Ingenierías (2019);11.

Saraf M, Pandya U, Thakkar A. Role of allelochemicals in plant growth promoting rhizobacteria for biocontrol of phytopathogens. *Microbiol Research* (2014) 169(1):18–29.

Shaheen S, Sundari KS. Exploring the applicability of PGPR to remediate residual organophosphate and carbamate pesticides used in agriculture fields. *International Journal of Agriculture Food Science & Technology* (2013) 4(10):947–954.

Sharma N, Singhvi R. (2017) Effects of Chemical Fertilizers and Pesticides on Human Health and Environment: A Review. *International Journal of Agriculture, Environment and Biotechnology* (2017) 10(6): 675-679.

Sheetal ML, Meena, Gehlot VS, Meena DC, Kishor S, Kishor S, Kumar S, Meena JK. Impact of biofertilizers on growth, yield and quality of tomato (*Lycopersicon esculentum* Mill.) cv. Pusa. *Journal of Pharmacognosy and Phytochemistry* (2017) 6(4): 1579-1583.

Shinwari KI, Shah AU, Afridi MI, Zeeshan M, Hussain H, Hussain J, Ahmad O. Application of plant growth promoting rhizobacteria in bioremediation of heavy metal polluted soil. *Asian Journal of Multidisciplinary Studies* (2015) 3(4): 179-185.

Simarmata T, Hersanti, Turmuktini T, Fitriatin BN, Setiawati MR, Purwanto. Application of Bioameliorant and Biofertilizers to Increase the Soil Health and Rice Productivity. *HAYATI Journal of Biosciences* (206) 23: 181-184.

Singh YK, Prasad VM, Singh SS and Singh RK. Effect of micronutrients and bio-fertilizers supplementation on growth, yield and quality of strawberry (*Fragaria × Ananassa* Duch.) cv. Chandler. *Journal of Multidisciplinary Advances in Research* (2015a) 4(1): 57-59.

Singh RP, Shelke GM, Kumar A, Jha PN. Biochemistry and genetics of ACC deaminase: a weapon to "stress ethylene" produced in plants. *Frontiers in Microbiology* (2015b) 6:1255

Singh M, Dotaniya M, Mishra A, Dotaniya CK, Regar KL, Lata M. Role of Biofertilizers in Conservation Agriculture. In: *Conservation Agriculture*, JK. Bisht et al., (eds.), Springer (2016) doi 10.1007/978-981-10-2558-7_4 (2016), 113-134.

Singh SK. Sustainable Agriculture: Biofertilizers withstanding Environmental Stress. *Int J Plant Anim Environ Sci* (2020) 10 (4): 158-178.

Thenmozhi R, Rejin K, Madhusudhanan K, Nagasathya A. Study on effectiveness of various bio-fertilizers on the growth & biomass production of selected vegetables. *Res J Agric Biol Sci* (2010) 6:296-301.

Verma JP, Yadav J, Tiwari KN, Lavakush SV. Impact of plant growth promoting rhizobacteria on crop production. *International Journal of Agricultural Research* (2010) 5:954–983.

Verma M, Sharma S, Prasad R. Liquid Biofertilizers: Advantages Over Carrier Based Biofertilizers for Sustainable Crop Production. *International Society of Environmental Botanists* (2011) 17(2): 16-18.

Verma P, Agrawal N, Shahi SK. Effect of rhizobacterial strain *Enterobacter cloacae* strain pglo9 on potato plant growth and yield. *Plant Arch (*2018) 18:2528–2532.

Vessey JK. Plant growth promoting Rhizobacteria as biofertilizers. *Journal of Plant and Soil* (2003) 25 (43): 511 – 586.

Wani SA, Chand S, Ali T. Potential use of *Azotobacter chroococcum* in crop production: an overview. *Current Agricultural Research Journal (*2013)1(1):35–38.

Yadav K, Sarkar S. Biofertilizers, Impact on Soil Fertility and Crop Productivity under Sustainable Agriculture (2018).

Youssef MM, Eissa MF. Biofertilizers and their role in management of plant parasitic nematodes. A review. E3 *Journal of Biotechnology and Pharmaceutical Research* (2014) 5(1):1–6.

Biographical Sketch

Bhupinder Dhir

Affiliation: *School of Sciences, Indira Gandhi National Open University, New Delhi, India.*

The Role of Biofertilizers in Sustainable Agriculture 201

Education: PhD Botany.

Business Address: E 25E, Vatika Apartments, Mayapuri, New Delhi - 110064, India.

Research and Professional Experience: My research experience is related to subject area of Environmental Biology, Environmental sciences, Plant Sciences, Agricultural Sciences and Toxicology. I have been working in the areas such as wastewater treatment, waste management, and development of eco-friendly technologies for wastewater treatment.

I have received many fellowships and successfully completed many research projects. The findings of research have been published as research papers in the research journals of national and international repute. I have also contributed to the field through publications in the form of review articles, book chapters and books.

Professional Appointments: I have worked as an Assistant Professor in reputed college of University of Delhi. I have received post-doctoral fellowships and research grants from government agencies. I have worked as an Educational consultant at Indira Gandhi National Open University, New Delhi. At present I am associated as a Faculty, Environmental Sciences at Guru Gobind Singh Indraprastha University, New Delhi.

Honors: Associated as an Editorial board member of Asian Council of Science editors.

Publications from the Last 3 Years:

Dhir B. (2022) Use of seaweed extracts for enhancement of crop production. In: Biostimulants for crop production and sustainable agriculture. *CAB International.* ISBN-10: 1789248078.

Dhir B. (2021) Biochars for remediation of recalcitrant soils to enhance agronomic performance. In: *Handbook on Assisted and Amendments Enhanced Sustainable Remediation Technology* John Wiley & Sons Ltd UK Majeti N. V. Prasad (Editor) ISBN: 978-1-119-67036-0.

Dhir B. (2021) Waste Water renovation; Wetland and Watershed Lake restoration. In: *Handbook of Ecological and Ecosystem Engineering* Majeti N. V. Prasad (Editor) John Wiley & Sons Ltd ISBN: 978-1-119-67853-3.

Dhir B. (2020) Plant Responses to Metalloid Accumulation. In: *Metalloid in Plants-Advances and Future prospects.* Deshmukh R., Tripathi D. K., Guerriero G. (eds.), Wiley.

Dhir B. (2020) Physiological basis of arsenic accumulation in aquatic plants. In: *Handbook of Bioremediation Physiological, Molecular and Biotechnological Interventions,* Academic Press, Pages 237-244.

Dhir B. (2019) Green technologies for the removal of agrochemicals by aquatic plants In: *Agrochemicals detection, treatment and remediation,* Elsevier, UK eBook ISBN: 9780081030189.

Dhir B. (2019) Removal of PPCPs by aquatic plants. In: *Pharmaceuticals and Personal Care Products: Waste Management and Treatment Technologies,* Elsevier, USA. ISBN 978-0-12-816189-0.

Index

A

abiotic, 2, 5, 10, 22, 25, 54, 56, 65, 73, 77, 86, 129, 137, 159, 185, 186, 188, 189

agriculture, vii, 1, 3, 4, 8, 9, 10, 15, 17, 19, 20, 21, 22, 24, 26, 27, 29, 30, 33, 34, 48, 52, 53, 55, 56, 57, 63, 82, 87, 94, 101, 107, 108, 110, 111, 112, 113, 117, 118, 119, 124, 130, 132, 134, 139, 140, 141, 150, 151, 154, 155, 156, 159, 160, 164,171, 173, 176, 177, 179, 195, 196, 197, 198, 199, 200

AM fungi, 1, 2, 3, 4, 5, 6, 7, 8, 9, 10, 11, 12, 13, 14, 15, 16, 17, 18, 19, 20, 21, 22, 23, 25, 27, 122, 131, 133, 189

application of AMF, 2

arbuscular, vii, 1, 4, 7, 20, 23, 24, 25, 26, 27, 28, 29, 30, 39, 43, 45, 46, 51, 52, 53, 57, 60, 61, 86, 100, 104, 109, 112, 114, 121, 131, 133, 134, 146, 148, 152, 155, 157, 183, 185, 188, 189, 190, 192, 193, 197, 199

artificial intelligence, 118

azotobacter, 11, 34, 41, 159, 161, 166, 167, 168, 169, 170, 171, 172, 173, 174, 175, 176, 177, 178, 181, 182, 185, 190, 191, 192, 200

azotobacteria, 160, 161, 166, 167, 168, 171, 173, 175, 176, 177

B

bacteria, vii, 3, 11, 16, 25, 26, 33, 35, 39, 41, 42, 43, 45, 46, 51, 52, 53, 54, 55, 56, 57, 59, 60, 61, 64, 65, 67, 68, 73, 77, 80, 83, 84, 86, 88, 89, 98, 100, 101, 109, 112, 121, 122, 124, 131, 132, 134, 139, 140, 143, 144, 145, 146, 147, 149, 152, 153, 154, 155, 156, 157, 159, 161, 163, 166, 167, 168, 170, 171, 172, 173, 174, 175, 176, 177, 178, 181, 182, 183, 184, 189, 190, 192, 193, 195, 196, 197, 198, 199

bacterial, vii, 17, 34, 42, 47, 49, 59, 63, 64, 66, 67, 69, 73, 78, 81, 82, 83, 84, 85, 87, 88, 89, 108, 111, 112, 114, 126, 129, 131, 144, 145, 147, 148, 149, 153, 156, 161, 166, 172, 173, 174, 175, 178, 183, 184, 188, 191, 196, 197, 198

beneficial, vii, 2, 3, 4, 5, 10, 11, 14, 16, 17, 21, 23, 34, 39, 43, 49, 50, 56, 57, 63, 65, 74, 80, 90, 100, 121, 122, 123, 125, 136, 137, 153, 154, 157, 159, 167, 170, 171, 175, 189

bioavailability, vii, 33, 35, 38, 40, 45, 47, 51, 55, 56, 93, 110, 113, 127, 139, 144, 154, 156, 199

biocontrol, 17, 23, 26, 30, 50, 64, 66, 71, 73, 74, 85, 86, 111, 115, 121, 123, 124, 125, 130, 131, 132, 135, 136, 187, 193, 199

biocontrol agent, 17, 23, 30, 64, 86, 111, 115, 124, 130, 132, 187

biofertilizer, 20, 24, 33, 34, 59, 60, 108, 113, 123, 153, 159, 160, 161, 171, 173, 175, 176, 178, 179, 180, 181, 184, 188, 192, 193, 196, 198, 199

biofertilizers, vii, 1, 2, 3, 4, 21, 25, 27, 34, 41, 45, 59, 112, 113, 130, 151, 156, 160, 166, 171, 173, 174, 175, 176, 177, 178, 179, 180, 181, 183, 184, 185, 186, 187,

204 Index

188, 189, 190, 191, 192, 193, 194, 195, 196, 197, 198, 199, 200
biopesticide, 64, 199
bioremediation, 63, 64, 66, 75, 76, 77, 78, 79, 80, 82, 83, 84, 89, 118, 120, 121, 124, 126, 127, 128, 130, 132, 136, 153, 154, 155, 156, 199, 202
biotechnologically, 117
BNF, 160, 161, 165, 168, 170
Burkholderia, vii, 34, 43, 44, 63, 64, 66, 67, 68, 69, 70, 71, 72, 73, 74, 75, 76, 77, 78, 79, 80, 81, 82, 83, 84, 85, 86, 87, 88, 89, 90, 91, 166, 167, 183, 184

C

chemical, 1, 3, 15, 21, 26, 27, 29, 33, 35, 36, 38, 40, 49, 52, 58, 67, 75, 78, 86, 95, 96, 97, 98, 99, 100, 101, 105, 107, 108, 109, 110, 111, 112, 114, 117, 118, 122, 123, 125, 130, 139, 141, 142, 159, 160, 161, 162, 168, 171, 174, 175, 179, 180, 182, 184, 193, 195, 196, 199
contaminated, 13, 30, 75, 77, 78, 83, 88, 118, 124, 126, 127, 128, 131, 132, 133, 134, 136, 137, 153, 155, 196
crop, vii, 1, 3, 4, 7, 8, 13, 15, 16, 17, 21, 23, 24, 28, 29, 34, 35, 37, 39, 48, 49, 50, 52, 58, 78, 82, 86, 93, 94, 95, 96, 97, 98, 99, 101, 104, 105, 107, 108, 109, 111, 114, 118, 122, 136, 139, 141, 151, 152, 153, 154, 159, 160, 161, 163, 164, 168, 170,175, 176, 177, 179, 180, 182, 191, 192, 193, 196, 197, 198, 199, 200, 201

D

deficiency, vii, 3, 33, 34, 35, 37, 38, 39, 40, 47, 48, 49, 50, 52, 53, 54, 55, 57, 58, 59, 60, 61, 107, 112, 129, 139, 141, 142, 143, 144, 155, 157, 163, 164, 165
diversity, vii, 19, 22, 24, 34, 36, 43, 56, 60, 68, 75, 82, 83, 88, 126, 145, 151, 155, 157, 160, 168, 169, 170, 175, 176

E

endosymbionts, 2
environment, 1, 2, 6, 12, 19, 23, 35, 48, 53, 67, 75, 77, 78, 80, 82, 93, 94, 96, 99, 107, 109, 110, 111, 113, 115, 117, 127, 129, 131, 132, 134, 139, 145, 146, 152, 154, 156, 159, 160, 166, 176, 179, 181, 193, 199
environmental, vii, 2, 3, 8, 9, 10, 15, 16, 17, 21, 23, 34, 35, 40, 45, 48, 63, 65, 67, 75, 77, 81, 82, 83, 84, 85, 86, 87, 88, 89, 90, 91, 93, 94, 101, 107, 108, 109, 111, 113, 117, 118, 120, 122, 123, 125, 130, 131, 135, 136, 150, 152, 160, 164, 165, 167, 168, 177, 179, 180, 194, 195, 197, 198, 199, 200, 201
enzymes, 18, 38, 51, 66, 73, 74, 75, 77, 83, 84, 123, 124, 127, 129, 131, 140, 143, 161, 166, 183, 185, 187, 189, 190

F

Fe solubilizing microbes, 140
fermentation processes, 118, 120, 122, 125
fertility, 3, 10, 26, 82, 90, 93, 94, 96, 104, 108, 109, 132, 150, 155, 159, 175, 176, 180, 191, 196, 199, 200
fertilization, 21, 54, 94, 107, 108, 112, 114, 118, 123, 126, 178, 196, 197
fixation, 17, 35, 63, 65, 68, 70, 73, 94, 134, 143, 155, 159, 161, 165, 166, 167, 170, 171, 175, 177, 179, 180, 182, 186, 187, 195, 198
functions, 24, 33, 34, 39, 54, 66, 77, 93, 123, 152, 153, 168, 187
fungal, 2, 4, 6, 8, 10, 11, 12, 13, 14, 15, 17, 18, 19, 20, 22, 23, 24, 26, 27, 29, 34, 57, 73, 74, 83, 85, 88, 115, 122, 123, 124, 125, 129, 132, 133, 145, 148, 149, 153, 182, 183, 185, 193
fungi, vii, 1, 3, 4, 5, 6, 7, 8, 9, 10, 11, 12, 13, 14, 15, 16, 17, 18, 19, 20, 21, 22, 23, 24, 25, 26, 27, 28, 29, 30, 31, 43, 44, 46, 51, 52, 53, 55, 56, 57, 58, 59, 60, 61, 64, 65, 71, 72, 84, 85, 86, 104, 108, 109,

Index

112, 121, 122, 124, 127, 131, 132, 133,134, 137, 139, 140, 141, 146, 147, 148, 149, 152, 155, 156, 157, 182, 183, 188, 190, 193, 197, 198

G

Glomus, 2, 5, 11, 19, 23, 25, 27, 29, 31, 45, 57, 60, 122, 133, 183, 188, 190, 192

I

inoculation, 2, 7, 8, 9, 10, 11, 12, 13, 14, 15, 16, 17, 20, 23, 24, 26, 28, 29, 30, 40, 41, 42, 45, 47, 51, 52, 55, 58, 59, 60, 69, 74, 84, 90, 91, 112, 122, 126, 131, 133, 135, 174, 175, 177, 188, 189, 190, 192, 196, 197, 199
insoluble phosphates, 118, 119
isolation, 2, 6, 29, 34, 41, 42, 50, 52, 53, 58, 59, 61, 85, 86, 87, 88, 89, 131, 144, 145, 156, 171, 173, 175, 176, 178, 198

M

mechanism(s), vii, 2, 4, 10, 12, 17, 18, 22, 24, 26, 27, 30, 34, 35, 36, 38, 39, 40, 43, 45, 46, 47, 48, 61, 63, 66, 68, 73, 76, 77, 78, 79, 81, 82, 86, 97, 108, 122, 124, 127, 132, 133, 139, 143, 147, 148, 149, 150, 151, 152, 153, 159, 170, 171, 175, 186, 187, 197, 198
metal, 3, 5, 21, 26, 46, 53, 54, 56, 57, 59, 76, 77, 78, 83, 86, 89, 124, 126, 127, 128, 132, 136, 137, 153, 184, 191, 196, 199
microbes, vii, 22, 24, 29, 34, 39, 40, 41, 43, 45, 46, 47, 49, 54, 57, 60, 75, 77, 88, 103, 104, 107, 111, 131, 139, 143, 145, 148, 150, 151, 152, 154, 157, 159, 174, 175, 181, 183, 187, 190, 191, 194, 196
microorganisms, 1, 3, 6, 13, 16, 17, 18, 21, 30, 33, 34, 36, 43, 48, 50, 75, 100, 102, 109, 110, 111, 113, 117, 118, 120, 121, 122, 124, 125, 126, 127, 128, 131, 135, 136, 137, 139, 140, 141, 143, 145, 146,

147, 148, 149, 150, 152, 153, 157, 165, 168, 171, 174, 180, 181, 184, 186, 194, 198, 199

N

nitrogen (N), 1, 3, 10, 12, 16, 28, 35, 39, 63, 65, 68, 70, 73, 81, 82, 84, 85, 86, 91, 94, 95, 103, 108, 110, 112, 114, 121, 122, 134, 143, 155, 159, 160, 161, 162, 163, 164, 165, 166, 167, 170, 171, 172, 173, 175, 176, 177, 179, 180, 181, 182, 186, 187, 189, 192, 193, 194, 195, 196, 198
nutrition, 5, 8, 11, 14, 22, 24, 26, 28, 29, 40, 49, 50, 51, 53, 55, 56, 58, 60, 87, 91, 94, 96, 97, 98, 99, 101, 105, 106, 107, 109, 110, 111, 114, 130, 133, 134, 137, 142, 152, 153, 154, 155, 157, 161, 165, 176, 177, 178, 186, 196, 198

O

organic acids, 34, 35, 39, 46, 99, 110, 120, 123, 124, 129, 133, 140, 144, 146, 147, 148, 149, 150, 151, 152, 165, 183, 184, 186
oxidizing, 184

P

pesticides, 15, 21, 27, 75, 77, 174, 180, 190, 199
PGPR, 3, 10, 11, 16, 17, 34, 35, 40, 41, 47, 58, 60, 67, 100, 134, 160, 177, 184, 185, 187, 188, 191, 196, 197, 198, 199
phosphate(s), 2, 10, 11, 16, 19, 22, 25, 26, 28, 34, 42, 50, 56, 57, 59, 63, 65, 70, 85, 95, 98, 99, 101, 102, 103, 104, 105, 108, 109, 110, 111, 112, 113, 114, 117, 118, 119, 120, 121, 122, 123, 124, 125, 126, 127, 128, 129, 131, 132, 133, 134, 135, 136, 137, 142,144, 147, 156, 159, 175, 178, 182, 183, 184, 186, 191, 192, 196, 198, 199

Index

phosphorus, vii, 1, 3, 5, 10, 24, 28, 29, 30, 35, 37, 45, 46, 49, 51, 54, 68, 82, 93, 94, 95, 97, 99, 100, 101, 102, 103, 104, 105, 106, 107, 108, 109, 110, 111, 112, 113, 114, 115, 118, 119, 125, 132, 133, 134, 137, 145, 152, 162, 173, 179, 180, 181, 183, 192, 195, 196, 197

phytopathogen, 64, 74, 87, 91, 123

plant growth, vii, 1, 3, 4, 5, 6, 7, 8, 10, 11, 12, 13, 14, 15, 16, 17, 21, 25, 26, 29, 35, 40, 41, 43, 45, 47, 48, 49, 50, 51, 52, 53, 54, 55, 57, 58, 59, 60, 63, 64, 66, 67, 68, 69, 77, 81, 82, 83, 87, 89, 91, 93, 94, 95, 96, 98, 99, 100, 101, 104, 111, 114, 118, 121, 123, 128, 131, 132, 133, 134, 135, 139, 140, 141, 144, 146, 147, 150, 151, 152, 153, 154, 155, 156, 157, 159, 161, 163, 164, 167, 170, 171, 175, 176, 177, 178, 179, 182, 184, 185, 186, 187, 188, 191, 193, 195, 196, 197, 198, 199, 200

plant growth promotion, 55, 57, 64, 66, 67, 153, 154, 171

promotions, 10

protection, 14, 56, 67, 82, 86, 111, 143, 166, 188, 189, 191, 194, 198

Q

quorum, 63, 64, 65, 66, 74, 82, 83, 90

quorum sensing, 63, 64, 65, 66, 82, 83, 90

R

resistance, 5, 15, 56, 67, 69, 73, 76, 77, 78, 81, 87, 124, 132, 185, 187, 189, 190, 191

rhizobacteria, 33, 35, 41, 45, 50, 52, 53, 55, 56, 58, 59, 82, 98, 100, 111, 113, 114, 132, 147, 154, 155, 156, 159, 177, 184, 195, 196, 197, 198, 199, 200

S

screening, 9, 30, 41, 42, 50, 56, 58, 59, 115, 144, 145, 171, 172, 175

secondary metabolites production, 64

secretions, 40, 74

siderophore, 18, 35, 40, 43, 45, 47, 52, 53, 68, 70, 77, 81, 82, 89, 91, 124, 140, 143, 152, 153, 155, 156, 171, 172, 191

soil fertility, vii, 1, 3, 7, 26, 34, 50, 94, 96, 97, 98, 100, 106, 107, 108, 109, 112, 125, 141, 150, 154, 155, 157, 160, 170, 175, 179, 191, 196, 198

stress, 10, 12, 21, 24, 25, 27, 39, 53, 54, 56, 58, 59, 60, 65, 69, 73, 76, 77, 79, 81, 82, 83, 87, 90, 91, 111, 128, 133, 134, 137, 156, 164, 179, 185, 188, 189, 191, 196, 197, 200

sulphur, 99, 113, 162, 180, 184

sustainable, vii, 1, 2, 3, 4, 8, 14, 15, 16, 18, 21, 22, 23, 24, 25, 26, 27, 28, 29, 33, 34, 36, 48, 49, 50, 56, 57, 63, 80, 93, 101, 107, 112, 113, 115, 118, 130, 136, 139, 146, 150, 151, 152, 155, 156, 157, 159, 160, 173, 175, 178, 179, 180, 195, 196, 197, 198, 199, 200, 201

sustainable agriculture, vii, 1, 3, 4, 14, 18, 21, 24, 27, 28, 33, 34, 36, 50, 56, 64, 107, 113, 139, 150, 151, 157, 159, 160, 178, 195, 196, 197, 199, 201

system, 6, 11, 27, 28, 39, 47, 53, 57, 67, 69, 74, 93, 95, 96, 101, 108, 109, 112, 123, 125, 127, 137, 148, 153, 159, 162, 175, 180, 185, 193, 196, 198

Z

zinc (Zn), vii, 3, 28, 30, 33, 34, 39, 42, 46, 47, 48, 49, 50, 51, 52, 53, 54, 55, 56, 57, 58, 59, 60, 61, 74, 77, 83, 86, 87, 89, 125, 126, 145, 153, 154, 162, 181, 184

Zn solubilization, 34, 42, 43, 45, 47, 54